Survival or Suicide

SURVIVAL

With chapters from the writings of

WILLIAM L. LAURENCE

DAVID E. LILIENTHAL

RAYMOND B. FOSDICK

HAROLD C. UREY

LELAND STOWE

JOHN FISCHER

EDWIN MULLER

QUENTIN REYNOLDS

HARRIS WOFFORD, JR. and others

OR SUICIDE

Editor, **HARRY H. MOORE**

A summons to old and young to build a united, peaceful world

Essay Index Reprint Series

 BOOKS FOR LIBRARIES PRESS
FREEPORT, NEW YORK

Copyright © 1948 by Harper & Brothers

All rights reserved

Reprinted 1971 by arrangement with
Harper and Row, Publishers, Inc.

INTERNATIONAL STANDARD BOOK NUMBER:
0-8369-2001-5

LIBRARY OF CONGRESS CATALOG CARD NUMBER:
77-134118

PRINTED IN THE UNITED STATES OF AMERICA

Contents

Preface ix

Part I
Which describes various problems and difficulties obstructing the peace

1. A New Era Ushers in a New and Greater Crisis 3
   WILLIAM L. LAURENCE
2. The Chaos of the Postwar World 8
   HARRY H. MOORE
3. Russia, the Master Problem 17
   JOHN FISCHER
4. The Problem of National Defense 24
   HARRY H. MOORE
5. War in the Atomic Era 31
   VARIOUS WRITERS
6. Getting Ready for World War III 46
   VARIOUS WRITERS

Part II
Which urges us to seek an understanding of the problems of the new era.

7. The Need for Understanding 55
   LELAND STOWE

vi	Contents

8	The Kind of Understanding Needed RAYMOND B. FOSDICK	60
9	War and Human Nature HARRY H. MOORE	66
10	The Problem of Understanding Russia JOHN FISCHER	74
11	The Larger Implications of Atomic Energy DAVID LILIENTHAL AND OTHERS	82
12	Training and Understanding for Youth HARRY H. MOORE	93

Part III
Which describes various ways of building a united, peaceful world

13	The United Nations HARRY H. MOORE	105
14	Proposals for World Government EDWIN MULLER HAROLD C. UREY AND OTHERS	115
15	Local Groups in Action HARRIS WOFFORD, JR. AND OTHERS	127
16	Beginnings of World-wide Friendliness QUENTIN REYNOLDS AND OTHERS	135
17	The Marshall Plan—An Adventure in Friendliness HARRY H. MOORE	143
18	Force and Friendliness HARRY H. MOORE	149

19	Mighty Enterprise	159
	HARRY H. MOORE	
	Appendices	
	A. Questions for Discussion	173
	B. Voluntary Organizations Promoting World Peace	178
	C. Educational Materials on War and Ways to Peace	185
	References	195
	Index	205

Preface

THE AIM OF THIS SMALL VOLUME IS TO EXPLAIN BRIEFLY the nature of the crisis brought about by the discovery of atomic energy, and to describe the momentous alternatives mankind now faces—a united, peaceful world achieved by the development of a strong international organization, or a third world war and the probable decay of western civilization.

The adequate development of this theme would be an audacious task for a single writer. To describe with accuracy and the requisite dramatic effect the epoch-making events at Los Alamos and Hiroshima; to understand and explain the baffling problem of Russia; to present satisfactorily the plight of Europe and Asia; to write authoritatively on the nature of war in the atomic age and on the causes of war in general; to set forth the larger implications of atomic energy in a manner commensurate with the far-reaching importance of the subject—all this requires background, facilities, and skills available only to specialists. It was believed, therefore, that a book containing the work of a group of such writers would have greater value than one prepared by a single author. The editor has undertaken to write particular chapters only because certain requisite material was not available in organized form. The result, then, is this somewhat unusual kind of anthology.

It consists (1) of chapters, or parts of chapters, taken from the works of outstanding authors which contribute to the aim of the book, and (2) of chapters the editor has prepared in order to fill gaps in the general plan of the volume and to provide a reasonable degree of completeness, unity, and continuity.

In the first part of the book, the chaos of the present postwar world is described, the menace of Russian communism, the apparent necessity of a program of strong national defense for the United States, and the tremendous costs, waste of wealth, and terrible risks inherent in such a program. Special attention is given to the nature of war in the atomic era. In Part II, emphasis is placed on the importance of our attaining an understanding of these and other problems brought about or accentuated by the discovery of atomic energy.

Then, in Part III, the work of the United Nations and the proposals now before the world for other forms of international organization are described. Efforts of various individuals and local groups in behalf of union and peace are here considered. Finally, the need for the leadership of the American people is set forth, and ways in which all of us—young and old—can help avert catastrophe and build a united, peaceful world. The choice before us is survival or suicide.

It should be especially pointed out that the book does not endorse any particular plan for international organization.

Preface

All proposals which have won a significant following have been presented for the consideration of the reader.

The book is not prepared for scientists or other intellectuals. For them, many scholarly volumes on various aspects of the present emergency are available. It is offered to those individuals—the "average layman" and also college and older high school students—who would welcome a brief, readable statement on the present crisis, but who have neither the time nor the inclination to pursue technical works. "Questions for Discussion," for the use of interested groups, may be found among the appendices.

The intelligent interest and the understanding of all, young and old, are essential. "Past thinking and methods did not prevent world wars," says Albert Einstein. "Future thinking *must* prevent wars." And we must carry the problems of atomic energy to the village square, he continues. "From there must come America's voice."

Marion M. Kingsley, as editorial assistant, has been indispensable in the building up of the book. She has examined a vast amount of periodical literature as well as various books, and she has done most of the condensing and most of the work on the appendices. Her judgment on general questions of policy has been very helpful.

Samuel Guy Inman, Maxwell Stewart and Donald Young have been especially generous in assisting the editor on various occasions. John Foster Dulles, Clyde Eagleton, Raymond B. Fosdick, George Franklin, Grayson Kirk, Marshall MacDuffie, Everet Minett, Fletcher Pratt, Arthur M. Squires, Robert Tilove and Edward R. Trapnell have kindly read particular chapters about which they have specialized or technical knowledge. Eric Berger, Adele Brebner and

Catherine Cleverdon have read all or parts of the manuscript and have made helpful suggestions. Various students have read parts of the manuscript. They include Herbert M. Jenkins, James Johnson, Ian MacKinnon, John A. Mellor, Grace Robson, Charles F. Steffens and Albert E. Tait, Jr. Kenneth M. Gould and Frances A. Thomas have been helpful in the selection of material for the appendices. David Baumann aided in the work of condensation, and Roger Ray gave many hours to critical reading and made valuable suggestions.

None of these various persons is in any way responsible for what has here been written. Furthermore, no contributing author is accountable for what the other authors have said or for what the editor has written. To the various authors and publishers whose material makes up a large part of the book special thanks are given; specific acknowledgments appear elsewhere. Whatever usefulness this volume may have will be due in large measure to the helpfulness and generosity both of advisers and of contributing authors.

In the chapters in the book made up largely or entirely of material from the work of other writers, the author's name, usually in brackets, may be found at the beginning of the material selected, and a footnote reference is given at the end; quotation marks are not employed. In other chapters, prepared by the editor, short passages, taken from the work of other authors, are indicated by quotation marks and references are given at the end of the book.

H. H. M.

Bronxville, N. Y.
April 8, 1948

PART I

Which describes various problems and difficulties obstructing the peace

You, at this moment, have the honor to belong to a generation, whose lips are touched by fire. . . . The human race now passes through one of its great crises. New ideas, new issues—a new call for men to carry on the work of righteousness, of charity, of courage, of patience, and of loyalty—all these things have come and are daily coming to you.

When you are old . . . however memory brings back this moment to your minds, let it be able to say to you: That was a great moment. It was the beginning of a new era. . . . This world in its crisis called for volunteers, for men of faith in life, of patience in service, of charity, and of insight. I responded to the call however I could. I volunteered to give myself to my master—the cause of humane and brave living. I studied, I loved, I labored, unsparingly and hopefully, to be worthy of my generation.

Josiah Royce

· 1 ·

A New Era Ushers in a New and Greater Crisis

It was at 3:25 p.m. on Wednesday, December 2, 1942, that a group of physicists, headed by Enrico Fermi and Leo Szilard, kindled the first atomic fire on earth. [William L. Laurence tells the story.]* On that day, in the gloomy squash court underneath the west stands of Stagg Field on the University of Chicago campus, man succeeded at last in operating a self-sustaining atomic furnace, the energy of which came from the vast cosmic reservoir supplying the sun and the stars with their radiant heat and light.

So formidable was the task that it is doubtful whether our scientists would have tackled it were it not for the news that trickled out of Germany that the Nazis had ordered their scientists to develop an atomic bomb. But early in 1942 Professor Arthur H. Compton telephoned Dr. Harvey C. Rentschler, Westinghouse research director. "How soon can Westinghouse supply three tons of metallic uranium?" Dr. Compton asked.

Dr. Rentschler was aghast. The total output of pure

---

* A statement regarding the use of material from the writings of various cooperating authors may be found in the Preface.

uranium metal up to that time had been a few grams. On being informed that uranium was necessary for a vital secret war project, he went to work. By November 7, 1942, a total of 12,400 pounds of pure uranium metal had been collected in the Chicago squash court, as well as tons of the purest graphite ever produced.

By the night of December 1, 1942, eleven layers of graphite uranium bricks had been piled up. Late that evening there were signs that the goal was near. Early the next morning the atomic "bricklayers" were back on the job. It was one of the coldest days of the winter. The squash court was badly heated but the bricklayers worked on, oblivious to the gloom and cold.

As the twelfth layer was completed everyone present became aware that one of the great moments in history was near. The instruments registering the multiplication rate of the neutrons (the number of neutron "bullets" produced for each neutron released in the pile) began clicking louder and louder. These clicks were the heralds of the atomic age.

The scientists had previously figured that if the number of neutrons per minute reached a count of more than 1,600 it would mean a multiplication factor greater than one, and the chain reaction would go on. Tensely and silently they stood around the neutron counters. Click, click, click. Twelve hundred, fourteen hundred, sixteen hundred. Sixteen hundred and one! The atomic age had come in on tiptoe.

While there was still much work ahead, the road was clear. It was a straight line from then on to New Mexico, Hiroshima, Nagasaki, and victory.

At Los Alamos, on the New Mexico desert, in July 1945, the atmosphere had grown tenser as the zero hour ap-

A New Era Ushers in a New and Greater Crisis 5

proached for the first trial of an atomic bomb. [Mr. Laurence was present and describes what happened.] It had grown cold in the desert, and many of us, lightly clad, shivered. Occasionally a drizzle came down, and the intermittent flashes of lightning made us turn apprehensive glances toward Zero, the tower on which the test bomb was set. We had had some disturbing reports that the test might be called off because of the weather. We knew there were two specially equipped B-29 Superfortresses high overhead to make observations and recordings in the upper atmosphere, but we could neither see nor hear them. We kept gazing through the blackness.

Suddenly, at 5:29, as we stood huddled around our radio, we heard a voice ringing through the darkness, sounding as though it had come from above the clouds: "Zero minus ten seconds!"

The voice from the clouds boomed out again: "Zero minus three seconds!" Silence reigned over the desert. We kept moving in small groups in the direction of Zero. From the east came the first faint signs of dawn.

And just at that instant there rose from the bowels of the earth a light not of this world, the light of many suns in one. It was a sunrise such as the world had never seen, a great green super-sun climbing in a fraction of a second to a height of more than eight thousand feet, rising ever higher until it touched the clouds, lighting up the earth and sky all around with a dazzling luminosity.

Up it went, a great ball of fire about a mile in diameter, changing colors as it kept shooting upward, from deep purple to orange, expanding, growing bigger, rising as it expanded, an elemental force freed from its bonds after being chained for billions of years. For a fleeting instant the color was unearthly green, such as one sees only in the corona of the sun during a total eclipse. One felt as though one were present at the moment of creation when God said: "Let there be light."

A great cloud rose from the ground and followed the trail of the great ball of fire. At first it was a giant column, which soon took the shape of a supramundane mushroom. For a fleeting instant it took the form of the Statue of Liberty magnified many times. Up it went, higher, higher, a giant mountain born in a few seconds instead of millions of years, quivering convulsively. It touched the multicolored clouds, pushed its summit through them, kept rising until it reached a height of 41,000 feet, 12,000 feet higher than the earth's highest mountain.

To another observer, Professor George B. Kistiakowsky of Harvard, the spectacle was "the nearest thing to doomsday that one could possibly imagine." "I am sure," he said, "that at the end of the world—in the last millisecond of the earth's existence—the last man will see what we have just seen!"

All through this very short but extremely long time-interval not a sound was heard. I could see the silhouettes of human forms motionless in little groups, like desert plants in the dark.

Then out of the great silence came a mighty thunder. For a brief interval the phenomena we had seen as light repeated themselves in terms of sound. It was like a blast from thousands of blockbusters going off simultaneously at one spot. The thunder reverberated all through the desert, bounced back and forth from the Sierra Oscuro range, echo upon echo. The ground trembled under our feet as in an earthquake.

The big boom came about one hundred seconds after

A New Era Ushers in a New and Greater Crisis

the great flash, the first cry of a newborn world—a newborn world with incalculable potentialities for good and for evil.*

Our new knowledge of atomic energy, combined with the understanding made available by a long series of inventions and discoveries since the beginning of the scientific revolution, may either save or ruin us.

We can use airplanes to carry serums for the cure of disease or to scatter disease germs. They are adapted to either purpose. We can use poison gases to exterminate the pests of civilization or to kill human beings. We can use atomic energy laboratories to produce isotopes for the conquest of cancer and other maladies, or to produce bombs to destroy civilization.

What will we do with our new knowledge and power? The question can no longer be evaded. For now we are confronted by world-wide chaos—starvation, disease, fear, economic depression, political upheavals, and the threat of a third world war. Survival or suicide? Those are the alternatives now confronting Western civilization.

Can man courageously face the crisis and develop the understanding and the moral attitudes which will enable him to utilize science constructively and build a world of abundance, health, culture, and good will among men?

There can be only one answer. We must build a united, peaceful world—or perish.

Old and young are summoned to this mighty enterprise.

---

* Reprinted from *Dawn Over Zero—The Story of the Atomic Bomb* by William L. Laurence, by permission of Alfred A. Knopf, Inc. Copyright, 1946, by William L. Laurence; and from three articles in the *New York Times* December 1, 1946, by William L. Laurence, used with permission of author and publisher.

· 2 ·

The Chaos of the Postwar World

THE CRISIS WHICH HAS BEEN SLOWLY DEVELOPING SINCE the inauguration of the scientific revolution became accentuated by the emergence of mechanized warfare in World War I. It became alarmingly acute with the discovery of atomic energy and the use of the atom bomb during World War II.

The war is over, but there is no peace. Approximately 17,500,000 men are at present under arms throughout the world; the military budgets of all nations add up to $30 billions.[1] Palestine, Greece, China and other countries are seething with conflict.

Chaos is now world-wide. Civilization and democracy in most of the major countries of the world have experienced a tragic decline. The war is over, but today more men, women and children are in political prisons and concentration camps than at any time in human history. In respect to the prisoners in Russian camps alone, estimates vary from 12,000,000 to 30,000,000. As late as December, 1947, in Great Britain, France, Belgium and other countries, there still were hundreds of thousands of war prisoners. Secret or political police (to be distinguished from legal or ordinary police) have grown with great rapidity—from five such police forces in 1939 to approximately 39 in 36 countries at present.[2]

Refugees and displaced persons constitute one of the most tragic by-products of the last war. They are people who have been uprooted from homes in Germany, Poland and other European countries, and the Far East. There are a million of them. A small number are willing to return to their native lands. The vast majority fear persecution because they are opposed to the new governments of their countries.[3]

The children of war-devastated and famine areas of the world are a major problem. There are at least two hundred millions of them.[4] Chester Bowles, early in 1948, returned from a survey of Europe. He had found, in an industrial town near Paris, in a window-less, one-room shack, an 11-year-old girl caring for three smaller children—and for her father, whose legs had been amputated. In Warsaw, in the corner of a cellar of a blasted house, were a boy of seven, his sister aged ten and their grandmother, who was earning a few cents an hour removing rubble by hand. The children were taking turns going to school, because there were not enough clothes for both.[5] There were innumerable instances of such conditions.

The hunger of the world is too great for most of us to comprehend. A summary appeared in the July, 1947, issue of *Survey Graphic*. "Most of the earth's population today can think of nothing beyond their desperate need for food —and their children's," pointed out Beulah Amidon, author of the article. "Millions of men, women and children are hungry—too hungry to work or to hope, dying of starvation or of the diseases that ravage the undernourished. This is why no problem in the world today is so urgent as that of food."[6] Then in August came reports of a European heat wave—in some regions the most intense in 50 years. The burning sun cut sharply into the yield of potatoes and late root crops. Soon a large part of Europe was in the throes of the worst drought in a decade. The result was greater hunger and more starvation than was earlier expected.[7]

Fortunately, the winter of 1947-1948 in Europe was a mild one, at least until February when a cold wave struck the continent. But shortages of food, according to a United Nations report, continued to be the world's number one economic problem.[8]

From country to country we find various degrees of devastation, depression and misery. Great Britain, early in 1948, was facing bankruptcy. The use of Marshall Plan funds, wrote Walter Lippmann, would make it possible only to mitigate the crisis. To prevent the catastrophe of mass unemployment and hunger, he said, would "require sacrifice, discipline and political audacity on a scale and in a degree which no free people in all history has ever had to show."[9]

In France, the crop failure, due to the drought, struck the people with stunning force. Sixty-nine infants per 1000, in early 1948, were dying within a year of birth—double the rate of the United States.[5]

In Italy, the average food consumption in February, 1948, was an enfeebling 1800 to 2000 calories per day (the average in America is 3300). Some families were going 24 hours without eating. In Hungary, some of the greatest physicians in the world were powerless to cope with the high incidence of malaria and tuberculosis, because of a tragic lack of medical supplies and hospital space.[5]

Poland, because she lost 65 per cent of her dairy herd

The Chaos of the Postwar World 11

during the war, may not become self-sufficient in milk for at least eight years. Her people are attempting to rebuild their nation almost with bare hands. Those who still live in Warsaw are huddled in cavelike rooms amid the wreckage. In Greece, many villages, during the winter of 1947-1948, were being deliberately and wantonly destroyed by Communist guerillas.[5] *

The misery of Asia, writes Foster Hailey of the *New York Times*, has been there for hundreds of years. But it has been "pyramided by war and neglect to one of the major problems of the world."[10] In India and Pakistan, where millions are always hungry, the late 1947 rice crop was estimated to be six per cent smaller than the previous year, and the wheat crop nine per cent smaller.

In China, starvation is widespread. While in 1946, in the remote famine area of Honan Province, children lay down in the streets and died, in late 1947, the dying were seen on the streets of modern Shanghai. Displaced by the Chinese civil war, millions of refugees have been flocking to nationalist-held towns and throwing themselves on the mercy of municipal administrations unable to support them.[11]

Thus, one may travel from country to country and find hunger the great pressing problem. All this, while we in the United States are eating more food than we did before the war.[12]

Finally, Japan's economy has been facing crisis; the United States may find itself saddled with a poorhouse

---

* It is true that Europe was increasing industrial production early in 1948 and may have been well on her way to recovery. But the clearing up of rubble, the mining of coal for larger exports, and the building of ships does not immediately provide improved shelter, clothing and food supplies. As late as February, 1948, there were still 60 million or more children and nursing and pregnant mothers in Europe and the Far East "actually starving or on the verge of starving." (See page 109.)

jammed with 80 million people. Economists have been divided over what to do.[13] In Germany, a new wave of hunger-strikes, in early 1948, swept through the merged American and British zones. Conditions were such that there was danger of complete collapse. The central fact in Germany is that, under the absolute rule of the victors, the country has been reduced to a slum and a penal colony. Germans have been forced to live on rations which often have approached those of the Nazi concentration camps.[14]

Abroad and at home, fascism is a danger of special importance in this time of crisis. "We defeated the Axis war machines," writes Leland Stowe.[15] "We did not crush fascism. . . . Bombs and arms alone can never [do that]. . . . Fascism gets its big chance from depression and accompanying mass unemployment. When people are without work, with little food, and with little hope, they are embittered. . . . They demand radical changes." In London and the northern industrial cities of England, wrote Drew Middleton in October, 1947, fascism and anti-Semitism were spreading. In Germany, Italy, and other countries, according to an authoritative report published in June, 1947, nazism and fascism "supported by considerable means and by steadily growing influences" were "openly starting to rise again."[16]

The future of South America also is menaced by the rise of the fascist movement. During the last war, fascist spies and Nazi-controlled enterprises functioned with impunity in Argentina. Franco's agents were equally favored. Fascism is now waiting for an opportunity to conquer the continent.

Fascism may become a serious danger in the United States. Between the two world wars, in periods of depres-

The Chaos of the Postwar World   13

sions and unrest, various fascist movements thrived here—the Silver Shirts, the German Bund, the White Shirts, and other groups. Now the Ku Klux Klan is coming to life again. "If you are a Catholic or a Jew or a foreign-born American," writes ex-governor Ellis Arnall of Georgia, "the cyclops and his twelve terrors may burn a fiery cross on your lawn some night and scare your children into convulsions. If you are a Negro, the hooded hoodlums may chase you for sport. And if you are a public official who feels that all citizens should enjoy equal protection of the law," the whole hierarchy of Wizards, Dragons, Cyclops, Kleagles, etc., "will harass you politically, threaten you with injury, and, perhaps, do a little casual conspiring against your life. ... If America does not get rid of them and of the ideology that underlies their nocturnal activities, they may get rid of America as we know it."[17]

Communism is now the major menace in Europe and Asia, as we shall presently see. There, our kind of capitalism no longer exists. But communism is spreading rapidly. The only alternative is social democracy, as recently developed in England and other European countries. It may go under, unless we in the United States support it by the contribution of money, food and clothing and, when essential, by military equipment and wise political aid.

The people of Europe are starved for ideas as well as for food, writes Ernest O. Hauser, an associate editor of the *Saturday Evening Post*,[18] after three years in Europe. "In our attempt to hold the line against Russia in Europe," he asserts, "we have not even begun to use ideological weapons." He writes of an all-day discussion with a top-ranking American business executive who was distressed at what he termed the "unconditional surrender" of the Western powers in the ideological field. At a World Youth Festival in Prague lasting four weeks in the summer of 1947, there were 13,000 foreign delegates. While 6,000 to 9,000 were communists, who were clearly attempting to win the support of the delegates, there were only 200 Americans.[19] A better understanding abroad may be developed, if the United States uses its expanded "Voice of America" program wisely and effectively.

The United States, notwithstanding its relative material prosperity, is not satisfactorily solving its various economic, social, and political problems. Our strikes, lock-outs, threats of depression, racial tensions, and economic injustice are not ignored abroad. The American and British zones in Germany were being flooded, in the summer of 1947, with Soviet propaganda leaflets denouncing democracy.[20] The Soviet press has published a report that the United States was dumping potatoes to keep prices up—a fact, of course; and the magazine *Krokodil* printed a cartoon showing a gloated plutocrat dumping a sack of large potatoes into a body of water, while a thin child looks on helplessly. Pictures of potato destruction have found their way, with the aid of the Communists, into the squares of Vienna and Paris and into the courtyards of Athens, as well as onto the collective farms of Russia.[21]

There are other symptoms of weakness in democracy here at home which can be readily observed by hundreds of Russian employees in the United States and reported to prospective followers in both satellite countries and those which may now be yielding to Russian pressure.

In our county jails, some hundreds of thousands of Americans are subjected each year to mistreatment almost as humiliating, unjust, and horrible as any uncovered in the concentration camps of enemy nations. Similar conditions are found in many of our mental hospitals. Court and grand-jury records contain evidence of "scores of deaths of patients following beatings by attendants."[22]

Our injustice to the Negro is a particular subject of critical comment in Europe. A poster showing a picture of a lynched Negro was exhibited at the World Youth Festival in Prague. It contained the caption, "Since V-J Day, 70 Negroes were lynched in the United States."[23] In many parts of the South, the Negro is not permitted to vote. Segregation in trains and other public vehicles is still operative in large portions of the United States.[24] The denial of franchise and other injustice to the Negro have been included in a documented bill of particulars submitted by the National Association for the Advancement of Colored People, late in 1947, to the United Nations.[25]

Finally, there is the fear that around the corner is another depression, which means unemployment, reduced earnings, child labor—poverty again in a land of plenty. Communist Russia is looking for such a depression and is expecting it. A great depression in "capitalistic United States" would be the signal for a further advance of communism throughout Europe and other parts of the world.

Employment reached an all-time high in 1947, business was excellent, and profits high. Large groups of workers had won reasonable wages, shorter hours, the comforts of life, and a new independence and dignity. Are capital and labor both willing to meet the conditions of continued prosperity? Have our leaders in industry and in government the necessary understanding of financial and economic problems?

These questions are of vast importance. While the people of Europe may react later against communism "by going fascist," there is greater danger now that they will not reject communism. Another depression in the United States now would strengthen communism. Recently the communists have taken advantage of American freedom, and through stealth and trickery have infiltrated the ranks of labor, the membership of various organizations promoting democracy, and even the staffs of several congressional committees.[26] A decline, even a slow decline, of democracy here may result in the breakdown of social democracy in Europe. Such a breakdown might make it possible for communism to win over the masses. A communistic Europe first, then a communistic South America, and a communistic Asia. And then, perhaps, the final tragic consummation.

Communist Russia is the master problem. To it we now turn our attention.

· 3 ·

Russia, the Master Problem

THE COLONEL STEPPED IN FRONT OF OUR CAR JUST AS IT swung onto the ramp which leads to the Forest Gate of the Kremlin. [John Fischer tells the story of his experiences and observations in Russia.] Three soldiers, carrying automatic rifles with bayonets fixed, moved out of the darkness just behind him. In the glare of the headlights I caught a glimpse of the bright blue collar tabs, edged with red, which mark the crack Internal Security Troops of the NKVD, the political police.

After a final careful scrutiny of passes, we were escorted through a maze of walnut-paneled anterooms and corridors to a little gallery overlooking the long hall where the Supreme Soviet—the Russian equivalent of Congress—was in session. Six plain-clothes men, looking like detectives the world over, lounged in the foyer which led to this gallery, and one of them accompanied us to our seats. He could hardly have regarded us—three American representatives of UNRRA with our interpreter—as dangerous characters; but he sat close behind us all the time we were there.

They conducted their deliberations under the eyes of blue-uniformed guards stationed at six-foot intervals along the wall. Additional guards were posted at each door, and at the end of every aisle stood an alert officer of the NKVD. I have never seen any building, not even Eisenhower's wartime headquarters, so elaborately protected.

It seemed a fair conclusion that somebody in the Kremlin was scared. A sense of fear quite plainly is one of the dominating facts of postwar Russia. But who is it that is so scared? And why?

The first answer is not hard to guess. The national fear neurosis could spring from only one source—the group of fourteen men who boss the Soviet Union. They make up the Politburo, the directing brain of the Communist Party. They also hold the key jobs controlling every nerve-strand of the country's administration. They are tough, able, and aggressive characters, and each of them controls a considerable span of power in his own right. What frightens them is not so easy to say. The scant facts at hand seem to indicate that the pattern of fear is a complex one. One element no doubt is sheer personal anxiety. The men behind the red wall remember how they rose to power, and they don't intend to leave that route open to any other group of determined conspirators.

One clue to Russian behaviour lies in the geography of the country. Naked plains stretch both east and west with no barriers to provide a defensible frontier. These plains have served as open highways for invading armies—Mongol, Tartar, Polish, Swedish, French, and German—since the earliest record of Slavic history. Fourteen times since 1800 hostile troops have poured across the western border; Minsk has suffered precisely 101 foreign occupations. And every invader since Genghis Khan has sent spies and fifth columnists ahead of his troops.

Russia, the Master Problem

The result is a suspicion of foreigners, a secretiveness, an obsession with security which was fully developed centuries before the present regime. The ancient obsession was, of course, greatly intensified by the shock of the last war. I have walked through cities such as Poltava and Kremenchug where 80 per cent of all buildings were razed. For two months I lived within half an hour's walk of the Baba Yar Ravine, where 140,000 bodies from the S.S. death camps were dumped in layers and covered by bulldozers. I have talked to nurses whose hospitals were soaked with gasoline and burned with the patients screaming in their beds. Nearly every family in the Ukraine has its own story of German terror. If these people view the outside world with a certain nervous mistrust, we may think it is regrettable, but we shouldn't be surprised.

Another reason for the Politburo's misgivings is its awareness of the country's enfeebled condition. Russia scraped through the war by the narrowest of margins—a margin provided only by Hitler's bad judgment and American Lend-Lease. During the course of the struggle it lost a stunningly high proportion of its factories, homes, farm machinery, livestock, and skilled workmen. Soviet officials told me frankly that they could not be completely replaced for at least a decade.

From the military standpoint, this means that Russia is like a giant temporarily exhausted from loss of blood. Even if the atom bomb had never been invented, the Soviet Union today and for some years to come cannot begin to match the military and industrial potential of the United States. And to the men in the Kremlin this is profoundly alarming.

This notion is understandable only in the peculiar terms

of modern Marxist theology. One of its basic articles of faith is the theory that the capitalist world can never escape from its fated cycle of booms and depressions, that each new depression is worse than the last, and that eventually the capitalist ruling class turns in desperation to fascism, imperialism, and aggressive war as the only way out of its economic troubles.

Consequently the Soviet Union must embark at once upon a new series of Five-Year Plans to raise its steel production and its output of other war materials. "Only under such conditions," the Generalissimo warned, "can we consider that our homeland will be guaranteed against all possible accidents."

Such "accidents" obviously are expected to originate in the United States, since it is the citadel of capitalism and the only remaining nation capable of challenging the Soviet power. The moment of danger, the communists believe, is likely to come in the decade after 1950, when they confidently expect America to sag into a catastrophic depression. According to Marxist doctrine, this almost certainly will result in a Hitler-like dictatorship, which will then embark on a campaign of imperialist aggression.

To the average Russian this warning seemed amply justified by Churchill's famous plan, voiced at Fulton, Missouri, for an Anglo-American alliance. The result was a shiver of horror through the entire country. Almost every day some item appeared about America's "imperialist" efforts to set up permanent military bases in Iceland and the Pacific. Professor P. F. Yudin has demanded a strengthening of the Red Army because "the Soviet Union is surrounded . . . by capitalist states which are constantly sending a stream of diversionists and spies."

Russia, the Master Problem

How far this text has sunk into the minds of the Russian people was illustrated by one typical seven-year old school girl. When an interpreter asked her what she wanted to be when she grew up, she replied: "A Red Army nurse to help fight the fascists who surround our country."

This fear of war is affecting the Russian people—their jobs, living standards, and personal freedom—and the direction in which it is pushing Soviet foreign policy. The Russians feel that they cannot safely put their trust in any world-wide association of nations so long as it might be dominated by what they regard as potential enemies. The Russians do not believe the United Nations would be any more effective in stopping an aggressive America than the League of Nations was capable of halting an aggressive Germany. They suspect that it might even turn into a grand alliance against the Soviet Union, because the Western democracies hold a majority in the UN and nearly always vote solidly together.

Consequently, the Soviet leaders are determined to organize their own security on quite a different pattern:

1. A strong Red Army, backed up by a war industry at least equal to any in the world.

2. A protective belt of satellite states, under firm Soviet control, outside of every vulnerable frontier.

3. Constant efforts to weaken and divide their potential enemies, the Western democracies, by every weapon of diplomacy and propaganda.

Here is not the One World we had hoped for but two; not the United Nations we had intended but two groups of nations which are anything but united. The Russian concept of security has split the world squarely into halves, and we are shut out of the Red half. We have had thrust upon us heavy responsibilities which we are reluctant to carry. When the world split in two, the United States inevitably became the center of one half, just as Russia became the center of the other. And all the nations outside the Soviet orbit automatically looked to us for leadership and support.

We have many reasons, then, for disliking this new pattern which the Russians have forced upon the world. There is no use pretending that we will ever like it. But in the end we are going to have to accept it because we have no other choice. So long as the Soviet leaders believe that an attack from the West is inevitable, they are not going to give up their kind of security.

No form of pressure can force them around to our way of thinking. Whether we like it or not, they have set the framework within which America must work out its own policy.

There is no prospect that the tugging contest can be ended soon, because the Russians believe that their protective ring of satellite countries is by no means complete. The underbelly of their vast double continent feels uncomfortably naked, as anyone can see by a glance at the map. Consequently, they would like to extend their security belt further to the south and east.

The clash of interests, therefore, will remain naked and unpleasant. But it is not likely to lead to an early war. The Russians will use every conceivable device of propaganda, political agitation, diplomatic pressure, and military threats to gain control over the areas they want. If these fail, however, they will not risk a major armed struggle. For at least fifteen years and perhaps longer they simply will not have the strength to challenge the Western democracies at any

Russia, the Master Problem

distance beyond their own borders. And by the time they have built up the power for such an attempt, there is at least a good chance of our persuading them that—if security is their only aim—further expansion is no longer necessary.

We can acknowledge frankly that it is not now possible to arrive at a sure and final answer to the question—What is Russia up to?—and we can then adopt a working hypothesis that will cover both possibilities. We can assume that the Russians probably are behaving in their present disturbing fashion because they are afraid; but that they may be getting ready to try to impose Communism on the rest of the world by force.

On an assumption of that kind, is it possible to build an American policy which holds some reasonable promise of peace? I think it is. It will be a long, tough, disagreeable job; it will be expensive; it will demand more patience and steadiness than we, as a nation, have ever shown before. But it should not be impossible.

To begin with, we'll have to recognize that nothing we can safely do at the moment will entirely dispel the Soviet fears. We must expect the present mistrust and tension to continue for a considerable period. During that period we must follow a line of action which will keep us strong against any possible attack and which at the same time will be calculated to prove to the Russians, eventually, that they have nothing to fear from us.*

---

* Condensed from *Why They Behave Like Russians*, by John Fischer, Harper & Brothers, New York, 1947.

· 4 ·

The Problem of National Defense

Believing that we cannot give up our monopoly on atom bombs, and that we must be prepared for the emergency that will suddenly arise when Russia can make atom bombs, the United States is adopting a policy of strong national defense. It is a policy that includes economic and military aid to Europe and Asia.*

Our immediate objective is to stop the spread of communism; and to this end we have already adopted various specific measures, primarily military. In addition, proposals for far-reaching developments, military and industrial, are now under consideration. The measures already taken include the following.

1. The maintenance of large appropriations for the armed services of the United States. The amount approved in the 1947-48 budget was $10,700,000,000.[1]
2. The passage, in May, 1947, of a $400,000,000 program for aid to Greece and Turkey.[2]
3. Provision for $275,000,000 of military aid to Greece and Turkey, and $125,000,000 to China, which can be used

* See also Chap. 17 on the Marshall Plan for economic aid to Europe and Asia.

The Problem of National Defense 25

for military purposes, in connection with the adoption, April 2, 1948, of the European Recovery Program.[3]

4. The acquisition of naval bases in Newfoundland and Bermuda, in Trinidad, in Georgetown, British Guiana, in the Midway Islands, in Guam, and in the Marianna Islands. These do not include naval bases in Point Barrow at the northern tip of Alaska and other bases in the Aleutian Islands and elsewhere.[4]

5. The negotiations of agreements with near-Eastern powers whereby large amounts of oil have become available to the United States for peacetime use or for war.

6. The maintenance and integration of an extensive intelligence service.

Proposals for measures which have received more or less support, but which, in early April, 1948, had not been officially adopted are:

1. The adoption of universal military training. President Truman's advisory commission estimated that this program will cost $1,750,000,000 per year and that between 750,000 and 950,000 boys and young men per year will be subject to this training.[5] At Fort Knox, Kentucky, the first of two or more experimental units for youths was set up early in 1947 as a sort of "pilot plant." It provided for 664 privates. If the Fort Knox Plan were adopted throughout the country, it might be difficult to maintain the high-grade supervision provided at Fort Knox. Hanson W. Baldwin, Military Editor of the New York *Times,* considers that the cost specified above (as estimated by the President's Commission) is definitely too low. His guess would be a cost between $2,500,000,000 and $3,500,000,000 per year.[6]

2. The temporary and limited resumption of the wartime draft. It was believed, in April, 1948, that about

3,600,000 non-veterans would thus be registered, of whom an estimated 1,355,000 would be eligible, after exemptions and deferments, for service.[7] Relatively few of these would probably be drafted.

3. A substantial increase in the production of military aircraft. The President's Air Policy Commission asserts that the minimum necessary at present is "an air force of 12,400 modern planes," supplemented by National Guard and Air Reserve groups.[8]*

4. An extension of the program of military research and development in aviation and related fields.[10]

5. The arming and training of armies and navies of other American republics at an annual cost of about $10 million,[11] the transfer of a maximum of 100 United States naval vessels, including a few cruisers, to South American countries for hemisphere security;[12] and the integration of the equipment of all armed forces on the two American continents with that of the armed services of the United States according to President Truman's message to Congress on May 26, 1947.[13]

* In presenting the Report of the President's Air Policy Commission, recommending a substantial expansion of air power, the Commission, explained Thomas K. Finletter its chairman, "did not like" the program which it recommended. "This is only a second-best way of doing things," he writes, "forced on us by the international situation and by the current fantastic developments in the military art—especially in the fields of atomic energy and biological agents, of aircraft and guided missiles. We said [in the report] that real security for this country lies only in the abolition of war, under a regime of world law, and that it is high time that the United States got busy to create such a world of peace . . . It is my faith that the number one foreign policy of the United States must be to establish world peace . . . [But] only a strong United States can have any chance of persuading the rest of the world to give up force . . . Our military establishment must be a shield behind which we can carry on a vigorous and positive policy for peace . . . Our military policy is, however, only the negative side of our foreign policy for peace. And by its nature it can be only a temporary policy. It will give us only a short time to do the positive things which will produce peace; for it is inflammatory. Time is running out, and we must make haste with our politics for peace."[9]

6. A budget of over $11,000,000,000 for the fiscal year 1948-49. Of this amount almost one-half is for the air force;[14] about $400,000,000 is for the operation of universal military training for the first year (included on the assumption that this measure will be adopted). The budget also includes about $550,000,000 for military research and development, exclusive of atomic research (the allotment for which is not specified) and about $285,000,000 for military stockpiling.[15]

7. A supplementary military budget, proposed April 1, 1948 by the President, of $3,000,000,000, to include the requirements of added military personnel, procurement of aircraft, aviation research and development, maintenance and operation, and various requirements and contingent expenses.[16] (The total amount first proposed by the Joint Chiefs of Staff, according to a writer in the *New York Times*, was $22,000,000,000. This was shaved down later to $9,500,000,000, and eventually reduced from "war" to "peace" proportions, in conference with the President to $3,000,000,000).[17]

8. A program for the decentralization of essential industries and the dwellings of the workers therein, together with the construction of underground caverns for industry. According to an estimate of J. Marschak, E. Teller, and L. R. Klein, of the Bulletin of Atomic Scientists, the total cost of decentralization alone would be about $300,000,000,000.[18] Furthermore, if and when such a program is actually launched, labor would probably have to be conscripted and it is unlikely that freedoms of speech and assembly could survive. Men would become mere instruments of the state in order to preserve the sovereign independence of the nation.[19]

Some Americans might have us go further in the development of a program of national defense than is here outlined. They would more boldly launch an extensive program of militant nationalism. There are a few who would even establish an American empire for purposes of defense. Outstanding among those who take this extreme position is James Burnham, Professor of Philosophy at New York University, whose book *The Struggle for the World* was published in the spring of 1947.[20] Few leaders of public opinion would fully endorse his proposals, but the book was greeted with extraordinary journalistic attention. *Time* regarded its appearance as an important news event, saying that "only one defense of Burnham's book can be made: it is—appallingly—true." *Life* devoted thirteen (partial) pages to a condensation of its contents.[21]

In brief, Professor Burnham believes that a third world war is already underway, that we are now engaged in the opening skirmishes—military as well as political and economic. The only possible way for us to win is to establish an American world empire in which this nation would maintain monopolistic control of all atomic weapons. Such a world empire should be established by concession where possible, says Burnham, but by force where necessary. If we do not establish an empire, communistic Russia will, and the peoples of the Western democracies would be reduced to slavery in communist totalitarian police states.

Professor Burnham knew that his frank proposal would shock a large part of the American people, but, he argues, "there is already an American empire, greatly expanded during these past five years." It takes in many of the islands of the Atlantic and most of the islands of the Pacific, "all the Americas," including Canada, and parts of Africa and Eu-

The Problem of National Defense 29

rope.* Communist Russia, believes Burnham, is moving toward the objective of world empire self-consciously and deliberately. Its leaders, he says, understand what is at stake. "But the Western power gropes and lurches. Few of its leaders even want to understand. Like an adolescent plunged into his first great moral problem, it wishes, above all, to avoid responsibility for choice. Genuine moral problems are, however, inescapable, and the refusal to make a choice is also a moral decision. . . . No wish or thought of ours can charm this issue away. This issue will be decided, and in our day. In the course of the decision, both of the present antagonists may, it is true, be destroyed. But one of them must be."[23]

Burnham apparently believes that the course of action he proposes would lead to the rolling back of Russia into a "contained" area, and that the United States would then have its "universal empire." He admits, however, that "the most hopeful route out of the crisis will be hard and painful and, most probably, bloody." It will be a struggle "such that one or the other, or perhaps both, of the contestants must in the end be defeated."[24]

The international crisis appears increasingly acute—even if the pessimistic dilemma set forth by Burnham is ignored. All of us want to stop the spread of communism. Apparently military strength is necessary for the present, whether or not one calls it "militant nationalism." Few persons,

---

* Territory held by the United States, according to the *New York Times*, includes Puerto Rico, the Virgin Islands, the Canal Zone, the Hawaiian Islands, Alaska, Guam, American Samoa, Wake, Midway, Canton, Enderbury, and the former Japanese Islands and mandates over which we have been given trusteeship by the United Nations.[22]

presumably, would endorse the establishment of an American empire.

The facts set forth in this chapter raise baffling questions. How far must we go toward militant nationalism? Is there any danger that present policy may lead to empire? Can we develop strong national defense as a temporary policy, while we seek to strengthen the United Nations or to build a world government as a permanent instrument to assure peace? If so, in which alternative, national defense or international cooperation, shall we invest our greater faith and the larger part of our resources? This is the major issue — to be set forth in our final chapters.

Now we must consider other questions —

Just how imminent is war? What would be its nature? What would it cost? What would be the outcome? We now turn to a consideration of these latter questions regarding war.

· 5 ·

War in the Atomic Era

THE ATOM BOMB IS NOT MERELY A NEW WEAPON: IT IS a revolution in war. [Thus spoke Dr. Philip Morrison before a committee of the United States Senate.] I saw the blackened ruins of Tokyo and Osaka, of Kobe and of Nagoya and I know that a city cannot live under the fire raids of a thousand B-29's. But the atomic bomb was something else. There were no shiploads of incendiaries in preparation for the attack on Hiroshima—no ordnance men and bomb dumps. There were about twenty-five people from Los Alamos, a few Quonset huts, and a barricaded building.

The strike took off about midnight. One plane with two escorts roared down the runway, took off, and set course for the enemy city. The reconnaissance photos next day told the story. One plane, with one bomb had destroyed many square miles of a city, destroyed them even more thoroughly and with even less chance for resistance or escape than the thousand-plane strikes.

Some time later we visited Hiroshima. I remember vividly the experience. The Japanese officials came there to talk to us and describe their experiences. I sat next to the chief medical officer of the district. He had been pinned in the wreckage of his house for several days after the explosion. He lived a little more than a mile from the point of impact.

Of 300 registered physicians, more than 260 were unable to aid the injured. Of 2400 nurses, orderlies, and trained first-aid workers, more than 1800 were made casualties in a single instant. It was the same everywhere. The military organization was destroyed; the commanding general and all his staff were killed, with some 5,000 soldiers of the garrison of 8,000. Not one hospital in the city was left in condition to shelter patients from the rain. The power and telephone service were both out over the whole central region of the city. There had been 33 modern fire stations in Hiroshima. Twenty-six were useless after the blast, and three-quarters of the firemen killed or missing. Debris filled the streets, and hundreds, even thousands, of fires burned unchecked among the injured and the dead. No one was able to fight them. Many places which had been only partly smashed by the blast were completely destroyed by fire.

Even more striking than the damage to buildings was the great number of casualties. Virtually all the people in the streets within almost a mile were instantly and seriously burned by the great heat of the bomb. These burns covered all the exposed flesh; sometimes even clothing caught fire and burned the wearer fatally. Many who escaped death from the blast or from burns died anyway. They died from the results of the radium-like rays emitted from the bomb at the instant of explosion.

In Nagasaki, which was struck by the second bomb a few days later, the homes are lightly built, but their factories are about like ours. There the Mitsubishi Torpedo Works buildings collapsed in a twisted jumble of steel onto the heads of the workmen and the still burning machines. For a good mile and a half all factory structures were totally destroyed.

It is likely that an American city would be as badly damaged as a Japanese city in an atomic raid, though it would look less wrecked from the air. In Japan the wreckage burned clean; in a Western city, the rubble would stand in piles in the streets. But the city would be just as ruined, and the people of the city just as dead.*

After the destruction of Hiroshima and Nagasaki the atomic scientists of the United States, alarmed by the terrifying power of atomic energy made available by their research, felt "a deep responsibility to tell every citizen what they know of the tremendous force—for good and evil—which they have released." Various groups of scientists and laymen have been organized for this purpose.

The Emergency Committee of Atomic Scientists, of which Albert Einstein is the chairman, has issued the following statement of facts "accepted by all scientists," which it believes should become known to the public.

1. Atomic bombs can now be made cheaply and in large number. They will become more destructive.

2. There is no military defense against atomic bombs, and none is to be expected.

3. Other nations can rediscover our secret processes by themselves.

4. Preparedness against atomic war is futile and, if attempted, will ruin the structure of our social order.

5. If war breaks out, atomic bombs will be used, and they will surely destroy our civilization.

* Abridged from the testimony of Dr. Philip Morrison before the Senate Committee on Atomic Energy, with minor revisions approved by Dr. Morrison.

6. There is no solution to this problem except international control of atomic energy and, ultimately, the elimination of war.[1]

A particularly ominous statement has been made by another group, the Atomic Scientists of Chicago. This organization asserts that a "time bomb of atomic explosive, smuggled into the country, can be introduced into a city ... in an automobile or light truck. ... Or, the 'makings' might be brought in piecemeal and assembled on the spot by an agent of an aggressor country."[2]

The facts agreed to by scientists have not yet reached a large proportion of the people of the United States. Many, to whom they have been given, have been skeptical as to their validity or have lacked the imagination to understand what the facts mean. Fortunately many meetings and radio discussions have been held, and considerable printed matter has been distributed in an effort to disseminate accurate information regarding the bomb and its uses.

At a University of Chicago Round Table in 1946, Harold C. Urey, Professor of Chemistry at the University, Edward A. Shils, Assistant Professor of Sociology at the same institution, and Thomas K. Finletter, formerly a special assistant to the Secretary of State, discussed the use of the atomic bomb. A part of their discussion follows.

Mr. Shils: But do you think that other nations will have the bomb?

Mr. Urey: Of course other nations will get the bomb in time. I suppose that in five, ten years—or maybe somewhat longer—there will be a great many atomic-energy plants scattered all around the world. They either may be for peaceful purposes or they may exist for the purpose of making war. But that will surely come and it will come regardless of "secrets." The matter of what atomic bombs will do has been discussed a great deal, but it is not possible, by the use of English words, to exaggerate the difficulties which have been brought into the world by the atomic bomb.

Mr. Finletter: I grant that, Urey. But, by your emphasis on the destructive power of atomic weapons, I take it that you are not minimizing the destructive power of other weapons—those which are already invented and those about which we are now beginning to talk.

For example, I saw in the paper the other day of various people appearing before one of the House committees and talking about the most fantastic kind of biological weapons. . . .

And I believe that General Marshall's report refers to a fifty-ton bomb already blueprinted by our ordnance, whereas the biggest bomb we used in this war was around two tons. That gives some indication of what is going to happen. . . .

Mr. Urey: If we live under the threat of total war of any kind, atomic bombs or otherwise, we are going to have to become a policed state in which we will lose all our liberties which we value so highly. Someone will have to tell us where we are going to live; whether we are going to scatter our cities or not; and so on.

Mr. Finletter: You mean to say that in your opinion, Urey, we will have a full authoritarian state, which will be bossed by a few people from the top and in which all our civil liberties will go. Is that correct?

Mr. Urey: I cannot imagine anything else.

Mr. Shils: Why will that be necessary?

Mr. Urey: It will be necessary because with the threat of

atomic bombs we have a new kind of warfare—one that will begin very suddenly and end in a very short time. Hence, it is necessary to keep on the alert at every minute, day and night, year in and year out. Otherwise, by means of the rockets carrying atomic bombs or by high-flying airplanes with atomic bombs, or perhaps by planting them in our cities—one way or another—bombs will come on a moment's notice without any declaration of war.

Mr. Shils: Then you think that we would have to be in a state of perpetual national emergency—a perpetual atomic alert. . . .

Mr. Urey: The time will come when some "Hitler" will arise in some country of the world and conclude wrongly (as Hitler concluded) that he can conquer the world and get for himself a great deal of power or a great advantage for his own country.

Mr. Finletter: Is it not true that these new weapons—the atomic bomb and all these other things—do lend themselves particularly to the advantage of the aggressor? In other words, the blitzkrieg seems more possible with these things; and, therefore, the aggressors are more apt to use them.

Mr. Shils: Is there not a danger, though, that we might even become aggressive ourselves in this terrible state of tension, not because we are deliberate or calculating about it, but because we will be so nervous, the tension will be so intolerable, that, in order to get the condition over with, some people will say, "Let's let the bombs fly?" And we will have a world war of atomic bombs.

Mr. Finletter: It seems to me that, with the kind of government which we have now—that is to say, a government which is run by the people—there is no possibility that the

War in the Atomic Era

people would stand for our starting on an aggressive war and throwing bombs around. But I am willing to concede that if we have the kind of authoritarian government which Urey has been talking about, even the American people might do anything, including starting a "preventive" war. I will concede that.

Mr. Shils: We should also make the point that people, because of the fear of destruction by atomic bombs and other new types of weapons, might become so irrational from panic in this country that they, out of sheer desperation, will say, "Let us have a war and get it over with, even if we are destroyed."*

Other new measures of destructive power to which Mr. Finletter alluded include (1) more powerful atomic bombs, (2) airplanes with greater speeds, (3) guided missiles, (4) incendiary bombs, (5) atomic radiation, and (6) biological warfare. Let us briefly consider the nature and effects of these measures.

1. *New atomic bombs*, in the opinion of Dr. Irving Langmuir of the General Electric Research Laboratory, may be "thousands of times more powerful" than the bombs used on Nagasaki and Hiroshima. Dr. Edward Teller of the University of Chicago has made a similar statement.[3] While the first bombs used cost approximately two billion dollars, bombs made later cost far less.

2. *Airplanes* will be available with far greater speeds than those ever before used in war. Douglas Aircraft has built the jet-propelled Skystreak which on August 25, 1947, flashed over the California desert at 650.6 miles per hour. Plane

* Condensed from the University of Chicago Round Table: The United Nations and the Bomb, Chicago, June 23, 1946. Reprinted by special permission of the University of Chicago Round Table.

manufacturers and the National Advisory Committee for Aeronautics are going full speed ahead on their research. Scientists are confident that we will have a fighter plane operating at a speed greater than that of sound (763 miles per hour).[4]

3. *Guided missiles,* according to Lieut. Colonel McCutcheon, may dominate World War III—if the war is staved off long enough. The V-weapons did enormous damage but they did not win the last war. They were inadequate because they could not be guided and controlled after launching. Now a few primitive guided missiles (such as the Navy's "Bat" and "Loon") have passed the blueprint stage. Colonel McCutcheon believes that their successors will have enormous ranges, striking at enemy cities and blasting them to rubble with atomic warheads.[5]

4. *Incendiary bombs* were used far more destructively during World War II than is generally known. On a night in March, 1945, they destroyed more Japanese property in a single raid than was wrecked by the Hiroshima and Nagasaki bombs combined. "Never before or since," the Army reported, "has so much destruction resulted from a single bombardment mission, regardless of the number of airplanes involved or the type of bombs employed." That mission, writes Milton Silverman, "turned nearly sixteen square miles of the industrial heart of Tokyo into rubble. . . . Even so-called fireproof buildings were completely gutted by the holocaust. More than a dozen of Japan's vital industrial targets were wiped out. Millions of people fled from the city, and traffic and communications were hopelessly disrupted."[6]

5. *Atomic radiation* presents new and ominous possibilities. At Hiroshima, many who escaped immediate death

succumbed within a few days or weeks to the effects of gamma rays, states a report by Ansley J. Coale for the Social Science Research Council. Those close to the bomb suffered from bloody diarrhea and expired within a week. Males, within a mile, suffered a loss of sperm cells, in some cases complete, for as long as three months. All known pregnant women within half a mile had miscarriages. A bomb dropped in the Hudson or Potomac Rivers might spray portions of New York or Washington (the column of water rises over a mile) with radioactive water whose deadly effect would be quite persistent.[7]

An atomic bomb planted by submarine, or dropped with a time fuse from a surface craft, into New York Harbor, asserts Lieutenant Colonel David B. Parker, could send several million tons of water a mile or more into the air. If a 30-mile wind were blowing toward the city, in less than an hour a deadly rain would be falling over the whole city. As a result, some hundreds of thousands of people might be killed, many by gamma radiation and others by being crushed to death in subway stations, on bridges, and in tunnels under the rivers. After the first hour of panic, the city's exits might be so clogged with wrecked cars and corpses that only a trickle of people, some of them swimmers, would be able to escape. Because of the difficulty in decontaminating buildings, the city might not be fit for repopulation by survivors for an entire year.[8]

6. *Poison gas warfare*, which may be utilized, promises to be inconceivably horrible. One finds it difficult to believe all reports. But according to one writer in the *United Nations World* there is a new gas called "Tabun" which drives men mad. It is said that one of the chief production centers is in the Russian zone of Germany. German scientists who

worked on poison gases are now in the Soviet Union, the United States, Britain and France.[9]

7. *Biological warfare* may prove to be the most deadly weapon of them all—deadly to crops, animals, and human beings. Considerable work has been done on plant killers, writes Professor Kenneth V. Thimann, Professor of Plant Physiology at Harvard University. Over one thousand related compounds have been made and tested. Some of these, spread at the rate of one to two pounds per acre, "will kill a wide variety of plants including many important crop plants. . . . Simple spraying is sufficient, so that large acreages could be destroyed efficiently from the air." Diseases of farm animals also can easily be spread.

Bacteria, deadly to human beings, continues Dr. Thimann, "could be sprayed on the enemy in various ways, in missiles or from the air. Cholera, dysentery, and bubonic plague would be obvious choices." A number of bacterial diseases are due to toxins which are probably the most virulent poisons known. A pound of diphtheria toxin, applied by injection, "could kill 13 million people." In practical use, it would be less effective than this. But if it is scattered, perhaps as a mist, as it might be, it would still be an exceedingly deadly poison. The toxin of botulism was isolated by the biological warfare group and its deadliness "is believed to be higher than that of the diphtheria toxin."

Preparatory work for biological warfare, concludes Professor Thimann, "does not require a Hanford or an Oak Ridge to manufacture major quantities of pathogens or their products." Furthermore, "in any system of international inspection, it would be difficult to trace either the personnel or the institutions involved."[10]

Thus, through the use of biological warfare, human beings

can be wiped out by the millions. There may be little civilian rubble, suggests Dr. Gerald Wendt, just corpses struck by poisons that kill swiftly and silently. A population could be wiped out without the slightest harm to buildings, docks, and transportation facilities. The enemy could take them over intact. In the next world-wide conflict, he concludes, germ attack may be a good way to start an undeclared war. While the ambassadors are still in fruitless negotiation, the enemy could decide there is no hope for peace and start uncontrollable epidemics simultaneously in many parts of the country. No great factories are needed, for the quantities of death-dealing materials are tiny compared with artillery ammunition. A small nation with sufficient talent could easily afford this kind of war.[11]

The United States is particularly vulnerable to bacterial warfare attack, according to the American Association of Scientific Workers. Our military experts are more apprehensive about Russia's use of bacterial weapons than their use of atomic bombs. In respect to the latter, we are still ahead; but in the development of germ warfare, the Russians are shoulder to shoulder with us.[12]

Should war come in this new atomic age, what would it cost—in terms of money, property, and human life?

An adequate estimate is obviously impossible. But, as a basis for an intelligent consideration of the question, let us examine the costs of the first two world wars.

In terms of money, World War I cost the United States over 22 billion dollars; the second world war, 330 billion dollars.[13] Our public debt soared so high during the last world war that the interest is now over 5 billion dollars per year.[14]

What is a billion dollars, just one billion? A billion one-dollar bills tightly stacked (not placed end to end) would reach 65 miles up into the stratosphere.[15] Compute for yourself how far 336 billion dollars would reach.

Now, for a moment, let us see what World War II cost us in terms of material wealth and institutions. Over 30,000 families could each be provided with an $8,000 house for the money the United States spent on the last war in *only one day*.[16] A permanent institution such as Columbia University, New York City, one of the largest in the world, could be established for the cost of one aircraft carrier, like the *Essex*, which soon becomes obsolete.[17]

In terms of the expense of killing one fighting man, it has been estimated that the cost of war has advanced as follows:[18]

Roman War against Gaul	$ .75
Napoleonic Wars	3,000.00
American Civil War	5,000.00
World War I	21,000.00
World War II	50,000.00

In terms of human life, what was the cost of World War II? Including soldiers and sailors who were killed in combat or died from battle wounds, the number of fatal casualties was approximately 10 million. This figure of course does not include the wounded and mutilated who recovered, the orphans, the widows, and the refugees. In addition, the number of civilians killed by bombings, execution, cruel treatment, or starvation is estimated to have been about 12 million.[19]

The figures on the cost of previous world wars are illuminating; but the atom bomb, as Dr. Philip Morrison

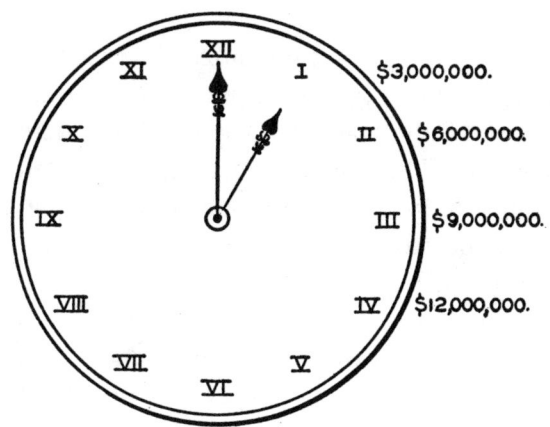

Figure 1[1]

said, has brought a revolution in warfare. Already we are considering the necessity of decentralizing industry and the populations of large cities, and of building new industrial establishments underground. Furthermore, in the next war (unlike the other world wars) the United States will almost certainly suffer from a vast destruction of houses, schools, libraries, art galleries, churches, and cathedrals, to say nothing of essential industrial and commercial establishments—if they are not moved underground soon enough.

Let us now briefly summarize. The cost of preparing for a new war now, at the beginning of the atomic age, and the cost of the war itself, may be considered under four heads: (1) The cost of preparedness or defense, (2) the cost in money of military operations during the war, (3) the cost of property destroyed, and (4) the cost in terms of human life.

The cost of national defense, as briefly set forth in the previous chapter, will obviously be enormous. The expense of a decentralization program alone, according to one estimate, will be almost as great as the cost to the United States of the entire second world war.*

The cost of military operations during the war may be relatively small, because the next war may not last more than a few days or hours. The first seven days, says General Dwight Eisenhower, will be decisive.[20]

The cost of property destroyed may be beyond all calculations. Consider what one bomb (now obsolete) did to Hiroshima. And remember that in the next war the United States will be for the first time a target (and a big target) for atomic bombs.

* See page 27, proposal 8.

War in the Atomic Era 45

The cost of human life may run into many millions. Ponder the effects of surprise attacks by new and more powerful atomic bombs, and the use of guided missiles, atomic radiation, and biological warfare. As already explained, entire populations of large cities may be quickly wiped out. If only atomic radiation and biological warfare are used, homes, factories, and cultural institutions may be left intact for the use of the enemy.

Is it conceivable that nations may permit themselves to be drawn into such a suicidal conflict as has here been described? We know that billions of dollars are being spent in preparation for war. But is this new and monstrous world war inevitable? Does the United States really expect this war?

We turn now to a consideration of recent developments which throw light on these questions.

· 6 ·

Getting Ready for World War III

BECAUSE OF THE POSSIBILITY THAT THE MEN IN THE Kremlin may "be planning an aggressive war to conquer the world for communism," to use the words of John Fischer, the United States is getting ready for such a war. There is space here to set forth only briefly some of the preparatory steps our government is now taking, and steps, also, that Russia is taking.

Emphasis is being placed on research. A small group of "highly skilled civilian engineers" is assisting air ordnance officers in reviewing hundreds of reports that may lead to new weapons.[1] In March, 1947, the National Inventors Council called on scientists, engineers, and amateur inventors to submit schemes for atomic-age warfare. "Practically any approach to the problem will be carefully considered."[2]

"An enormous amount of basic research," says the magazine *Time*, will be required. "During the war, the world used up its accumulation of unutilized discoveries in physics. . . . Now the nation's research reservoir will have to be filled again. This is being done. Government funds

Getting Ready for World War III

are pouring into universities and special foundations for laboratories packed with costly apparatus."[3]

In the United States and seventeen other countries, in 1947, atomic energy research was underway with approximately 125,000 scientists participating at an annual expense of about $500,000,000.[4]

Industrial preparedness against the contingency of a third world war was the keystone of government planning in Washington early in 1947 [writes Donald B. Robinson]. Top-ranking military men believe that the one way to victory, as well as the best hope of avoiding renewed war, is to keep the nation's industrial war machine in a state of immediate readiness for emergency. Below are the main points of Robinson's statement.

The best brains of the Army, the Navy, and private industry are drafting an over-all secret industrial mobilization plan to keep American mills and factories geared for instant conversion to war production. They call for careful cataloguing of every mill and factory in the country capable of conversion to war production.

An Army and Navy Munitions Board is setting up some seventy industrial advisory committees to deal with all phases of governmental and industrial war activity. Army and Navy procurement officers are in the field inspecting potential war plants. The part each of these plants will play in national defense, the weapons it will make, and even the schedule upon which it will turn them out are being determined now.

The Army is also anxious for legislation which would provide for a labor draft in case of World War III. Overtures are now being made to the CIO and the AFL to win organized labor's support for such a measure.

The menace of the atomic bomb has made the study of underground plants an A-1 priority duty for the Board. Army men have studied England's enormous mile-long underground factory which came unscarred through the entire Battle of Britain. They have reviewed Japanese techniques in building underground shelters, hangars, and storage facilities. The Army is believed to have obtained reports that Russia, with prisoner-of-war labor, has built the biggest underground airfield in the world. It is said to be located "in a highly dominating and commanding area which has a great, semi-global air striking potential."

It is said that, contrary to expectations, caves have not been found generally suitable for manufacturing or storage. Ordinary hillsides may be preferable to mines and quarries. Eventually, indicated Kenneth C. Royall, when Undersecretary of War, the government may have to force industry to move plants away from such built-up sections as Detroit and Pittsburgh and reconstruct them underground or in hillsides.*

In World War III, there will possibly be another Pearl Harbor, but it may be closer to our homes. Our military planners frankly expect an attack from the north. Canada is generously cooperating with us in providing defenses of our northern frontier.

The military planners [writes William B. Arthur] keep much of their work secret. But their goal is apparent: they want a defense line running from Okinawa through

---

* Condensed from *The Army's Plan for the Next War*, by Donald B. Robinson, *The American Mercury*, February, 1947. Used with the permission of the author and publisher.

Mr. Royall, on August 3, 1947, revealed plans for continued government control of 131 war plants. He also indicated that a comprehensive survey of underground sites had been completed.[5]—Editor.

Getting Ready for World War III

Alaska and across the northern plains of Canada, with a possible eastern anchor in Greenland or Iceland. Following are other developments explained by Mr. Arthur.

Our forces are not asleep—now. There in the Arctic regions, they know what they face. And they are doing something about it. In February, 1946, the Navy carried out "Operation Frostbite." A task force, including the giant aircraft carrier *Midway*, pushed into Davis Strait between Labrador and Greenland. It worked its way up to within 200 miles of the Arctic Circle. It proved that floating aircraft bases could be used in the cold regions.

In December, 1946, our Army ground forces were maneuvering in "Operation Frigid." In sub-zero cold, 1700 officers and men of a special unit were operating over Alaskan terrain. The Army was also conducting "Operation Williwaw" around Adak, in the Aleutian chain. Both forces were testing tanks, self-propelled guns, special snow vehicles, radar, arctic kitchens, rations, and clothing.

A sharp spur to this planning is Russian activity in the Far North. Soviet experts have been carrying on polar experiments for years.* And the Russians have started a five-year hydrographic plan in their northern waters. Some 450 expeditions were to set up lighthouses, radio beacons, radar stations.

But there are more ominous signs. The Russians are building a gigantic military base at Petropavlovsk. This is on Siberia's Kamchatka Peninsula, only 490 miles from Attu in the Aleutians, and only 4910 miles from Detroit's huge factories.

* The Soviet recruiting drive for German and other European physicists and engineers, asserted Joseph and Stewart Alsop, July 10, 1947, "has been without parallel, and Soviet orders now overload" Swiss, Swedish, and other European factories.[6]—Editor.

There they are throwing up acre after acre of warehouses and barracks around Petropavlovsk's harbor, its airdrome, and back of the city. Throughout northern Russia and Siberia the soviets have set up an intricate system of ice-scouting by airplane. They are sending planes out over the Aleutians from time to time, but call them routine weather patrol flights.*

Will all the new weapons be used in World War III? Is it possible, for instance, that the horrors of atomic radiation and biological warfare will be perpetrated on the human race? Apparently the University of Chicago thinks there is real danger. Late in 1947 it began instruction on what to do for victims of atomic war. The subjects being studied include the effect on human tissue of death rays liberated by atomic bombs.[7]

At a gathering of prominent chemists in March, 1947, John M. Hancock, an associate of Bernard M. Baruch in formulating the United States plan for the control of atomic energy, may have given the correct answer to this disturbing question. He declared, "No matter what the nations agree to as to use of any tools of war, and no matter to what high level the moral conscience of the world has risen in this regard, those are likely to vanish if once a war starts—for that is the ultimate breakdown of any rules of conduct among nations. At that time the right of self-defense, the need to survive, will govern the conduct of men and nations."[8]

I am a frightened man [declares Dr. Harold C. Urey].

* Condensed from "We're Getting Ready for War in Alaska," by William B. Arthur, Look, December 10, 1946, p. 31. Used with permission of the author and publisher.

Getting Ready for World War III

All the scientists I know are frightened—frightened for their lives—and frightened for your life.

I say to you—and I wish I could say it face to face—that we who have lived for years in the shadow of the atomic bomb are well acquainted with fear, and it is a fear you should share if we are intelligently to meet our problems.

The world has become much smaller and much more explosive.

Now we are all crowded together into a single house. Beneath the floor of our house there is a time bomb ticking away, as I write this . . . as you read this. Nations are crowded into a very small space as considered by the standards of the supersonic rocket and the atomic age. The explosion of an atomic war would smash our house of civilization—smash it beyond human comprehension. Those who even think of an atomic arms race, those who boast of battleships and air power, those who speak of using national force to maintain peace, simply do not understand this crowded house of fear.

They do not think of what the situation may be a few years from now.

Atomic war could unleash forces of evil so strong no power of good could stop them. Make no mistake. Other civilizations have died because they would not learn their lessons in time. Remember that if Hitler had beaten us to this weapon—as he beat us to the V-2—America today would be a slave province in a fascist World State.

What we would not learn from Hitler we must learn from Hiroshima!

Other issues wait. Other problems will stand delay. But

the main race between man's powers for evil and his powers for good—that race is close to a decision. The bomb is fused. The time is short.

We must think fast. We must think straight.*

---

* Condensed from "I'm a Frightened Man," by Harold C. Urey, as told to Michael Amrine. Copyright, 1946, The Crowell-Collier Publishing Co. Used with permission of the author and publisher.

PART II

Which urges us to seek an understanding of the problems of the new era

AFTER HAVING CREATED THIS ATOMIC AGE, AMERICAN education cannot escape the responsibility for teaching men how to live in it without destroying themselves. . . . More than ships, planes and guns, we need an intelligent and well-informed public opinion. . . . The power of democracy is the power of uncensored knowledge, of unregimented minds, of resolute action based on a realistic understanding of a realistic world.

OMAR N. BRADLEY[1]

· 7 ·

The Need for Understanding

WHEN OUR SECRETARY OF STATE PROPOSED THE "Marshall Plan," he made it clear that, because of our strength and resources, we must assume leadership in the building of the peace. Are we prepared to assume such leadership? Mr. Leland Stowe, journalist, gives one answer in this chapter.

The most frightening thing in today's world [says Mr. Stowe] is the terrible unpreparedness of the American people either to cooperate constructively for peace or to assume their necessary role in world leadership.

This psychological and educational unpreparedness is almost as frightening as the menace of the atomic bomb. In one respect it is possibly more dangerous. We can all sense and see the peril of the bomb. But can we see, do we understand, the menace which lies in ourselves? The fact is plain that the world we now know cannot survive for long unless it is better managed than the old world ever was. The fact is plain that *neither men nor governments can safely manage what they do not understand.*

The American people have lost a great deal of their

keen, intelligent interest in political trends and political meanings which characterized so many of our national leaders between 1776 and 1865. Americans have tended to become self-satisfied or timid and indifferent in their political thinking while becoming engrossed in technical skills and mechanical progress; we have become increasingly less interested in the functionings of democracy while becoming more and more interested in the functioning of turbines and physical production.

While yet we have not succeeded in establishing a tolerable brotherhood among our own citizens, destiny has elected us to exert decisive leadership toward a brotherhood of mankind and a peace capable of saving the world's foremost nations from destruction.

We must see with new eyes the society in which we live, and the world in which we live. Somehow we must make this part of our daily business. We must understand that the fate of a Welsh coal miner, a French or Hungarian shopkeeper, a Greek or Chinese peasant, a Russian collective farmer, and a Mississippi Negro without a vote —all this is the fate of you and me. We may turn our backs upon this fate, but we shall never escape it.

Would you venture to claim that we Americans are adequately prepared for world leadership? Indeed, have we even been educated for peace? The Office of Public Opinion Research at Princeton uncovered the following facts: Some 27,000,000 (or nearly one out of three among 90,000,000 adults in the U.S.A.) did not know that the Japanese occupied the Philippines. When asked whether the United States had ever been a member of the League of Nations, 30 per cent replied in the affirmative—and another 26 per cent did not know. Only 32 per cent could

The Need for Understanding

say how a treaty is approved in this country. Facts like these are difficult to square with the proud boast of "the best educational system in the world." They also raise an important question. If between 30 and 50 per cent of American adults do not, and will not, read their daily press sufficiently to pass a fair current-events test for twelve-year olds, how are they going to get the knowledge on which to vote or act intelligently upon the supreme problems of war-or-survival which now beset us all?

On the basis of the facts which are all around us, it is difficult indeed to be an optimist about a world in which we Americans exert the decisive balance of power. Most of us do not read much more than the headlines about what is happening in this revolutionary world. Only the most biased and sensational of our daily press are read by the millions. The pulp magazines, the "true stories," the "love stories," the "western," the "detective," the "movie star" publications—all so expertly designed for the paralyzing of thought—are read by tens of millions of our adults. The atomic time-fuse means that human beings, of all nations, have never before needed so urgently to inform themselves—and *to think*. Instead of this, for the most part, the movies and the radio are drugging us with entertainment.

We must look sharply at our entire American system of education. For generations and from the first grade through high school (at the very least) we have been educating our children for war; and this we are still doing in a blissful sort of blindness. Admittedly, the school systems of most other countries have done precisely the same thing. Under the United Nations a totally new emphasis on mutual understanding and cooperation could—and should

—be injected into a large proportion of the world's school rooms. If we Americans, however, really wish to avoid the disaster of an atomic war, we would do well to take the initiative within our own borders. The place to begin building peace is at home.

When I say that we are unconsciously educating for war, I mean things we can put our fingers on. Nationalism that promotes a feeling of "better than thou" and "everything American is best" is education for war. Any nationalism that is arrogantly self-assertive and scornful of other peoples' way of life is a mental conditioning to "put other nations in their place." We are by nature such a patriotic people, that we have not realized the extremes to which we have been carried. This pronounced American nationalism once whipped us overnight into a war with hapless Spain.

For hundreds of years, nationalism of this kind has produced nothing but wars. Our new world cannot exist very much longer with this kind of education. To educate for peace we must clean the slate, begin anew, and begin at home. American taxpayers paid more than $2,000,000,000 for the atomic bomb. Suppose we were now to spend another $2,000,000,000 for a new and modernized nation-wide system of education? In one generation this might well give us far more protection from mass destruction than 100,000 atomic bombs can provide. First of all, it is unseeing minds and shriveled hearts which produce war. For if there is to be any peace, it is people—and only people—who can achieve it.

At the close of a hearing of the U.S. Senate's Special Committee on Atomic Energy the following dialogue occurred:

Senator Johnson: "I have one further observation to make, and that is that you scientists have got a long way ahead of human conduct; and until human conduct catches up with you, we are in a precarious condition—unless you scientists slow up a little and let us catch up."

Dr. Langmuir: "Scientists are not going to slow up. They are going faster."

Senator Johnson: "Then, we will have to speed up."

Dr. Langmuir: "You will have to speed up."*

---

* Reprinted from *While Time Remains* by Leland Stowe, with permission of Alfred A. Knopf, Inc. Copyright, 1946, by Alfred A. Knopf, Inc.

· 8 ·

The Kind of Understanding Needed

We "will have to speed up." Surely, Dr. Langmuir is right. And so is Dr. Urey right. "We must think fast," he said. "We must think straight."

Today, as we face the task of building a united world, we must, through laborious effort, seek understanding. We must have superior minds now for the solution of our social, economic, and international problems. Raymond B. Fosdick understands the subject thoroughly, and we present in this chapter a condensed statement of his firm belief.

Our great need for brains [declares Mr. Fosdick] is not in the natural sciences, but in the whole field of the social sciences. The pursuit of truth, the hunger for knowledge have at last led us to the tools by which we can ourselves become the destroyers of our own institutions and all the bright hopes of the race. Man is confronted by the tragic irony that when he has been most successful in pushing out the boundaries of knowledge, he has most endangered the chance of human life on this planet.

In this situation what are we to do? Shall we curb natural

science and try to fix boundaries beyond which intellectual adventure shall not be allowed to go? This is impossible. The search for truth is the noblest expression of the human spirit. The unconquerable urge for knowledge about ourselves, our environment, the forces by which we are surrounded, the universe in which we live—this is the spark which gives human life its meaning and purpose and clothes it with final dignity.

No, we cannot put brakes on intellectual adventure. But there is a lack of balance about our studies and our research which cries for correction. We are discovering the right things but in the wrong order, which is another way of saying that we are learning how to control nature before we have learned how to control ourselves. The disproportion between the physical power at our disposal and our capacity to make good use of it is growing with every day that passes.

There is developing a dangerously tilted situation in our society, an intellectual imbalance, which we can no longer ignore. Our knowledge of human behavior and social relations is not adequate to give us the guidance we need; and the fundamental issue of our time is whether we can develop understanding and wisdom reliable enough to serve as a chart in working out the problems of human relations, or whether we shall allow our present lopsided progress to develop to a point that capsizes our civilization in a catastrophe of immeasurable proportions.

International trade and finance, national income and its distribution, wages, profits, prices, purchasing power, employment and unemployment, collective bargaining, housing, crime, population, agriculture, transportation, and the social, economic, and political organizations that deal with these matters—these are a few topics, selected at random, about which we must have deeper and more adequate knowledge if our society is to be kept in equilibrium. Russia has an economy of scarcity; the United States has an economy of abundance. What does this mean in terms of world unity? What are its implications? How can the two economies be reconciled so that they can live in peace together on the same planet?

The problems included in this kind of research are for a variety of reasons far more involved and complex than the problems which the natural scientists are facing. Social issues cannot be clearly defined and understood except on the foundation of hard, painstaking work. We must have disciplined minds and the high integrity of objective scholarship; and the flow of first-class talent into these fields must be continuous and uninterrupted.

Senator Kilgore said recently that we must have "a sufficient mastery of nature so that permanent world peace will be a reality and not a mere hopeful expression of faith." With due respect to the Senator, it is this mastery of nature which threatens to blow our civilization into drifting dust. What we really need is a mastery of man's social nature—knowledge and more knowledge of the onrushing social consequences of our machines, consequences which, because they are too intricate to be easily understood, are shaping our lives to ends we do not want but cannot escape.

In the social sciences we need boldness, a spirit of daring, a certain scorn of the past, a fearless facing of facts. The present emergency requires a fundamental reappraisal of things that have hitherto been regarded as more or less sacrosanct. It calls for an atmosphere of hospitality to new

The Kind of Understanding Needed 63

ideas, of open-mindedness to the work of those pioneers who are digging in the almost unexplored fields of the social sciences.

Surely in any list of social institutions and ideas that requires analysis and restatement, the current conception of patriotism would stand well at the top. What is this thing we call patriotism? Once a noble passion that broke down local provincialisms and stretched the mind to broader loyalties, today, with the expansion of international life, its tendency is to narrow rather than widen the sympathies of men. Once patriotism was a unifying force that brought order among small conflicting groups; today, in the world-wide society of mankind it has become a disintegrating force.

In this word patriotism, therefore, we have a conception that needs critical analysis and reappraisal. Here is an opportunity for the integrity and detachment of the laboratory. Here is a chance to dissect from a great human virtue the malevolent growth that is devouring it—the hundred-per cent Americanism, the flag-waving nationalism, and all the tribal self-infatuation and arrogance that mask behind man's instinct for loyalty.

The aims and something of the mood of physics and chemistry are beginning to influence the newer sciences of man. Although the subject matter is far more intricate and unmanageable, here and there an attempt is being made to carry over to the social sciences the inductive technique and quantitative method of natural science. At many points research into human nature, human beliefs, and human institutions is being eagerly promoted.

Fortunately, collaboration and cooperation between all types of science, rather than the intensification of old rivalries, are increasingly evident as diverse disciplines like economics and agriculture and the biological sciences and physics are harnessed together to meet particular human problems. We are accustomed to terms like biophysics and biochemistry; but today we have a lengthening list that includes psychobiology, medical economics, legal medicine, social psychology, psychosomatic medicine.

The development of a new anesthetic, for example, involves the chemist, the pharmacologist, the pathologist, the neurophysiologist, the surgeon, the statistician, and the psychiatrist. The elimination of malaria from a particular region brings together into a single team ecology, entomology, epidemiology, immunology, parasitology, botany, limnology, engineering, chemistry, political science, law, economics, and sociology.

Indeed, during the war it was necessary for governmental agencies to develop mixed teams of physical, biological, and social scientists in what came to be known as "operational research"—a type that was geared not to a particular discipline but to the problem that had to be met.

Whether the problem was national defense or the fighting power of the armed forces or public health or making an atomic bomb, it immediately overflowed the boundaries of the natural sciences into the area of social relations, so that in the end there were no boundaries, no lines of separation. Instead there was a vast single problem which had to be met with all the tools of knowledge that were available.

This is the direction in which the tide is moving; this is what the necessities of the present emergency dictate. The world of scholarship is not two worlds; it is one world.

It is not a segment of the truth that will make us free;

The Kind of Understanding Needed

it is all the truth—the whole truth, inclusive, interfused, unified, guided by the principle that while a real knowledge of man is impossible without a knowledge of nature, a knowledge of nature is sterile unless it is linked with a wider knowledge of man.*

A wider knowledge of man evidently must be attained—a better understanding of human nature, particularly as it is related to war and peace. To this problem we now turn our attention.

---

* From *The Old Savage in the New Civilization*, by Raymond B. Fosdick, copyright, 1928, by Doubleday, Doran & Company, Inc.; and *We are Living in Two Centuries*, by the same author, in the *New York Times Magazine*, Nov. 24, 1946. Minor revisions have been approved by the author. Used with the permission of the Doubleday Co. and the *New York Times*.

· 9 ·

War and Human Nature

Before the first world war, more than 160 organizations were working to promote peace. By the time the second world war began such efforts had increased enormously. Eight peace organizations had annual budgets in 1939 ranging from $50,000 to over $500,000 each. Although expenditures for peace ran into the millions, and although these endeavors had able leaders and the support of thousands of devoted persons, yet they appear to have come to naught.

One great difficulty which still retards peace is that people in general, and our statesmen in particular, do not have a thorough understanding of human nature and its relationship to war and peace. We do not know why men fight.

While it is often said that it is human nature to fight, psychologists do not agree. Some 2,400 of them have declared almost unanimously, "War is not born in men; it is built into men. No race, nation, or social group is inevitably warlike."

Why then do we have wars? It is easy to explain a defensive war. When the United States was attacked at

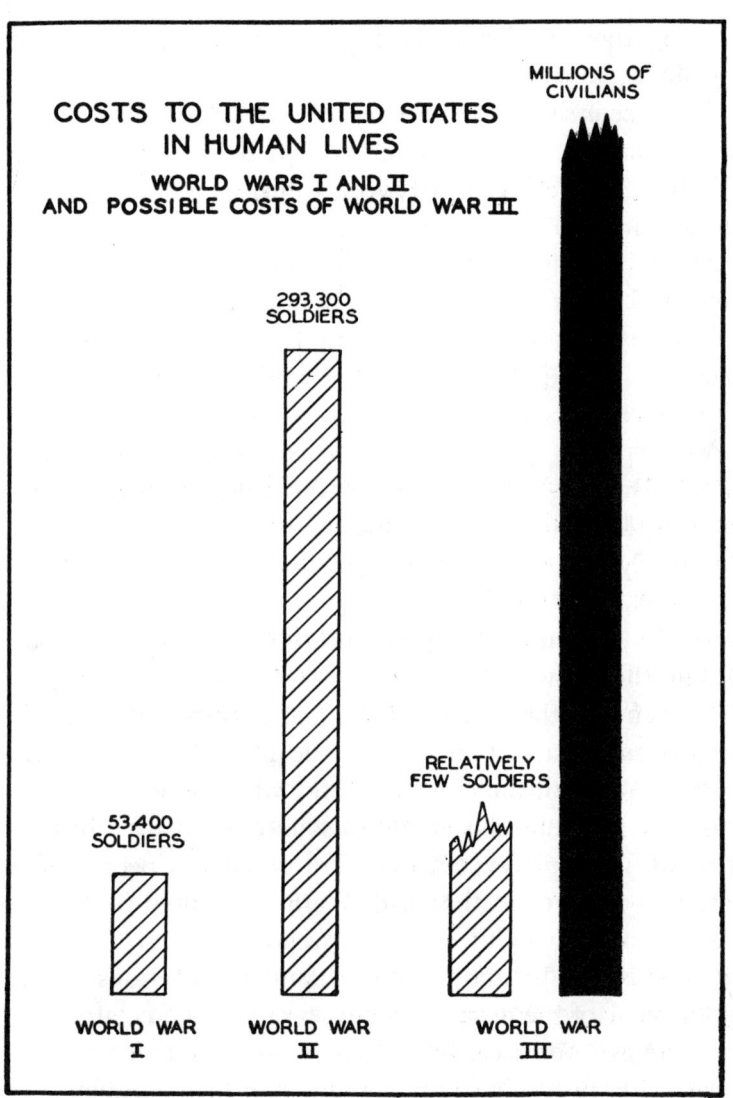

Figure 2[2]

Pearl Harbor, we fought back. The difficult problem is to explain aggressive warfare.

Injustice is said to be one cause of war. Yet history shows that attempts to correct injustice can be a cause for war. We have long thought that economic causes play an important part in bringing on war, that the "have-not" nations are the aggressors. But Switzerland, Denmark, and Peru and other nations are "have-not" nations in respect to various natural resources, and China and India have been the hungriest people of the world, yet have not engaged in aggressive warfare against other nations.

War springs fundamentally out of rivalries for power, asserts Quincy Wright, not out of economic rivalries, although the latter may develop as a result of the former. But surely this is not the full explanation.

Finally, it is said that it is primarily the malcontents who are responsible for war—those who are disturbed within themselves. "People are most warlike and aggressive when (1) the social and economic relationships of life have been disrupted so that people feel bewildered, confused, uncertain, and insecure; and when (2) people are frustrated." While this may be a sound explanation of war, it is not a simple explanation. For when the people of a nation today are frustrated, they are "frustrated in a complex economic, political, and psychological setting." All factors need to be understood in their interrelations before a clear view of the origins and causes of war can be attained.

There is "one principle upon which psychologists can agree with full conviction," writes Gardner Murphy. "It is that peace cannot be achieved as a negative goal, the absence of something at which we shudder." The problem then, he says, is to find a "substitute for war's disciplinary function," to preserve as far as possible the martial virtues such as courage, self-sacrifice, duty, and endurance, without the awful cruelty and wastefulness of war. James thought that this could be done in the collective enterprises of the future, in youthful armies "enlisted against Nature," in required periods of work in factory and field.

Various substitutes have been suggested by psychologists. "In the development of the arts lies the ultimate hope for true peace," asserts Laurence K. Frank. "There is only one instrument for channeling the intense emotions of a people: the arts, especially drama, which has historically provided the catharsis of strong emotions."

"In many respects strenuous athletic rivalries present, better than modern military service, the conditions for which militarists argue," wrote Walter B. Cannon, physiologist, in 1929. "Man is pitted against man, and all the resources of the body are summoned in the eager struggle for victory. In the Philippine Island . . . there were no athletics before the American occupation. The fierce Igarots of Bontoc, once constantly at war with neighboring tribes, now show their prowess not in head-hunting, but in baseball, wrestling and the tug-of-war."

"We who have known the pioneer life in countries like America and New Zealand," says Walter Nash, former New Zealand Ambassador to the United States, "know how fully the natural instincts of combat, excitement and comradeship can be satisfied in struggle against the elements. We know that there is as much satisfaction to be found in such a contest as in any struggle against a human foe. Any man who has worked on, shall we say, a great electric power scheme . . . knows full well that peace, no less than war, can be crowded with exciting efforts—

with heroic achievement—effort and achievement moreover, which bring no suffering and destruction in their train but a happier and better life for all mankind."

The United States, during the depression of the early thirties, established the Civilian Conservation Corps. Camps of youths, scattered throughout the country, fought forest fires, built trails and roads in national and state forests, improved national parks, drained swamps, and planted trees and built dams for the prevention of soil erosion.

The enormous tasks of postwar reconstruction also furnish a promising substitute for war, suggests Houston Peterson. "The Arctic and Antarctic, the desert and the jungle, are waiting to be conquered, as well as the ocean and the air. It may be felt that these are for the more active and gifted of youth; but it is precisely with them that we are chiefly concerned. And each may participate in accordance with his ability and his will, once the leadership is given. With such fields for heroism, discipline and self-sacrifice opened up, further organized killing may well seem anti-climactic and superfluous."

"There remains," writes Dr. Peterson, "that universal and ceaseless war against disease. Infantile paralysis, tuberculosis, cancer, syphilis and all their cohorts surpass in deadliness and ruthlessness any human enemy. When will the masters of psychological warfare begin to dramatize this greatest of all conflicts? We are prepared to fight infantile paralysis with the planned strategy of a military campaign, said President Roosevelt to the nation on his birthday, January 30, 1943. The gallant chapter workers, the doctors and nurses in our hospitals, the public-health officials, the volunteers who go into epidemic areas to help the physicians—these are the front-line fighters."

War and Human Nature

Thus we see that the problem is vast. Just now it is the major problem of mankind. The sciences of man, particularly the psychological sciences, must be mobilized to find a solution. A good beginning has been made, but far more research is essential. Gardner Murphy emphasizes its importance. Human nature must be subjected to far more scientific study, he asserts, with special reference to international strains and dangers. Research in this field must become as international as physics. There should be exchange fellowships and exchange teachers with funds for laboratory, statistical, and field surveys. Public opinion must be studied more thoroughly. Research centers are necessary to receive data from local and national groups and to funnel to such groups the results of studies throughout the world.*

Are the various opinions of psychologists, set forth here, important to leaders of public opinion, educators, legislators, and governmental administrators, particularly to statesmen in national and international bodies who are striving to prevent war and establish peace?

Surely, if such understanding as we have of human nature and war could be more generally disseminated, all of us, eager for peace, could work more intelligently. Psychologists and the other social scientists have a definite contribution to make to current deliberations on peace. Why should they not be invited to participate? Should not the United Nations and our own government set up

---

* There are, in addition, long-term problems related to the prevention of war and the development of a world of peace—especially the problems of food supply (including soil erosion) and the growth of population. These intricate, long-term problems cannot be satisfactorily discussed in a small volume such as this.

commissions to study the causes and the prevention of war from the point of view of psychology?

Is it important that the role of drama, athletics, pioneering, and the fight against disease be considered *now*, when every day brings us closer to a decisive crisis? Is there time, one might ask, for such "theorizing"? Yes, if the psychologists are right in asserting that peace cannot be attained as a negative goal. A Hitler cries for blood, we are told, not an Einstein. Accumulated resentments and other disturbing emotions now in evidence must be drained off and drained off promptly. It is not enough that food for the body be provided. A nation with a program of constructive activities such as those proposed, is not so likely to need destructive outlets for repressed emotions. And let that nation also enlist the energies of its youth in the strenuous intense struggle against floods, forest fires, swamps, impenetrable forests, and similar obstacles to human welfare. The over-all problem, it has been emphasized, "is the democratic one of making life far more interesting, varied, dignified for far more people."

Can we be assured that these various observations and conclusions are sound? Many hundreds of psychologists and other social scientists would agree that they are. Surely they are entitled to careful study.

Greater understanding can be attained through further research. Adequate facilities and funds for investigations in this field might bring us closer to world peace than research directed toward the building of larger bombs.*

---

* This chapter is based almost entirely on *Human Nature and Enduring Peace*, the Third Yearbook of the Society for the Psychological Study of Social Issues, published for Reynal & Hitchcock, Inc. by Houghton Mifflin Company, Boston, 1945; and on *War and Human Nature*, by Sylvanus M. Duvall, Pamphlet no. 125 of the Public Affairs Committee, 1947; with

A better understanding of human nature and war may throw light on the present warlike behaviour of the Russians and on our difficulty in getting along with them. This practical problem now deserves our attention.

---

the permission of the Society for the Psychological Study of Social Issues and the Public Affairs Committee.

The paragraph by Walter B. Cannon is a condensed statement from his book *Bodily Changes in Pain, Hunger, Fear and Rage*, D. Appleton & Co., New York, 1929, pp. 387, 388.

· 10 ·

The Problem of Understanding Russia

It is one of the basic tenets of Marxism that the capitalist world can find no way out of its crisis except through imperialism and war. [John Fischer now continues his discussion of Russia.] That is why Lenin once predicted that "a series of terrible conflicts between the Soviet Republic and the bourgeois states is unavoidable." Nothing we can say is ever likely to budge them an inch. For the trained communist has his own system of logic, which he believes to be infallible. Anyone who does not speak in the same terms is heard with kindly pity.

There is only one way, it seems to me, to cure the Russians of their tragic apprehensions—and it will take a long while. We simply need to pull through the next fifteen years without a major depression and without going "fascist." If we can find some democratic method of controlling the violent ups and downs of our economy—if we can hold onto full employment and our freedom at the same time—then we will have proved beyond question that the communist forebodings are all wrong. That kind of proof is the only sort Stalin and his associates will readily accept.

The Problem of Understanding Russia 75

When that time comes—and, I think, not until then—the Soviets may be convinced that there is a possibility of real cooperation with the non-communist world.* They may begin to rely for security on the United Nations, rather than the Red Army, and to ease up their frantic efforts to build a huge defense industry. The Kremlin might then feel that a further expansion of its security zone is unnecessary, and that it can afford to loosen the lead-strings on its present satellite countries. It might even feel safe in risking some measure of democracy and free speech within its own borders.

But in the meantime we cannot afford the kind of wavering and ambiguous policy which left Germany in doubt about our willingness to defend Europe. We need to make it perfectly clear that we are committed to defend certain vital areas, that we will fight if they are invaded, and that we have the strength to fight successfully. If we draw that sort of line, we can be quite certain that the Red Army will not cross it.

But this is not simply a military problem. America's future strength will be largely determined by our success in holding our lead both in industrial capacity and in the techniques of production. That means a healthy, stable economy. From a strictly military standpoint, we cannot afford another depression.

A depression in this country would mean catastrophe for most of our allies. It would spell unemployment and short rations for British factory hands, closed doors for French innkeepers, bankruptcy for Brazilian coffee grow-

---

* There may then be a change in Russian tactics, says Stefan Osusky, visiting professor at Colgate University, toward a practical and complete collaboration.[1] See also page 156.—Editor.

ers, probably starvation for Bolivian tin miners and Indian jute farmers. It would be a death sentence for democracy throughout wide stretches of the Western community; hungry people notoriously fall easy victims to the infections of either communism or fascism.

In the future, United States foreign policy will no longer rest in the hands of Washington. It will depend on the energy, foresight, and responsible behavior of the whole body of American citizens. Every investor, corporation director, and trade union leader will be making foreign policy in the day-to-day handling of his business. And if our economic leaders are really serious in their opposition to communism, they will waste no time in getting together to figure out means of avoiding another depression.

There are also weak spots in our security zone which have to be shored up promptly. The men in the Kremlin have seized every opportunity to pose as the champions of the hungry, hopeless, and oppressed everywhere from Manchuria to Athens. Their propaganda promises a new deal: steady jobs, a burst of economic development, freedom from the old foreign bosses or native feudal rulers —and it points to the remarkable industrial progress of the Soviet Union as proof. There is only one way we can answer this challenge. We must demonstrate that our system will provide a better living and more freedom than the communists can offer.

We can offer more. We might well supply the capital, the engineers, and the equipment for a Jordan River authority in Palestine, for example, and a Tigris and Euphrates River Authority in the Arab lands just to the east. (Herbert Hoover has already set forth in some detail just how the job could be done.) American machinery and know-how could save twenty years—the crucial twenty years—in the long-overdue industrialization of India and China. And incidentally, programs of this kind, on a really large scale, would take up a lot of the slack when our domestic demand for capital equipment begins to taper off in three or four years.

All this will cost money, of course; being a great power never has come cheap. It will provoke grumbling among our more myopic politicians about giving a quart of milk to every Hottentot. But in time, if they are not completely blind, they will come to see that we have to deliver the milk—or the Soviets will. Our economic strength is the greatest single advantage we have in the contest against the communist half of the world; I am not yet ready to believe that we are fools enough not to use it where it will do the most good.

In the military field, Soviet officials feel—as a number of them explained to me—that they have had several legitimate reasons for worry. They see us with a navy greater than all other navies in the world combined; an air force more powerful than all others combined; the only really long-range bombers in the world. Knowing their own temporary weakness, they see no power capable of launching an attack against the United States for years to come. And yet they see us spending more money on armaments than any nation in the world, and far more than we ever before spent in peacetime.

It is perfectly plain that no scheme of general disarmament has the slightest chance of working unless it provides for an airtight system of international inspection and control, not only for atomic energy but for all other weapons of mass destruction.

The best we can do is to keep on trying, with bottomless patience, to persuade the Russians to accept a workable system of disarmament under international control. There is at least a chance that eventually—once the boundaries between East and West have become reasonably stabilized—they will come around.

Finally, in our day-to-day dealings with the Russians we need to learn an almost superhuman forbearance and tact. We must remember that they are Orientals with the same bargaining instincts as an Armenian rug dealer. To them it is normal to start out with an asking price ten times as high as the price they finally expect to close for. They also expect the customer to walk out the door a few times while the haggling is going on. They simply can't understand our taking offense at such tactics, and they can't believe us when we say that we cannot recede from a position taken at the beginning of negotiations.

Such people obviously are going to be difficult to get along with; but we must learn to do it, with all the tolerance and patience we can muster. The American diplomat whose opinion on Russian affairs I value most once privately summed up the problem in these terms:

"We can contribute only by a long-term policy of firmness, patience, and understanding, designed to keep the Russians confronted with superior strength at every juncture where they might otherwise be inclined to encroach upon the vital interests of a stable and peaceful world; but we should do this in so friendly and unprovocative a manner that its basic purposes will not be subject to misinterpretation."

Four or five years from now the division of the world between the communists and the Western democracies probably will be fairly well completed. We may then find

The Problem of Understanding Russia

a more stable world than these gloomy and turbulent times lead us to expect.

For the first time, the two greatest powers will be widely separated, rather than face to face across a fortified border or narrow seas. War between them would appear to be less likely, simply because either one would find it almost impossible to reach the vital centers of the other. The atom bomb and the long-range guided missiles will not change this fact. We and the Russians might destroy each other's cities in an atomic cataclysm, but neither could gain decisive victory without occupying the territory of the other.

Once the present adjustments to the new two-power arrangement have been completed, therefore, we might reasonably look forward to a considerable period of stability. It would hardly be accurate to call it peace; it will more nearly resemble an armed truce. But if the balance can be held even, if the Western half of the world can remain prosperous, strong and democratic—then over the course of years the men in the Kremlin may get over their fearfulness and aggressiveness. And eventually—perhaps in another generation—the truce may be converted into peace.*

Since these words were written, the speeding up of the Communists' time table in Europe seems to suggest, not that it is important to "understand" the Russians and cure them "of their fear of foreign attack," but that our big task now is to stop them.

Nevertheless, if we are to proceed intelligently, it is necessary that we understand Russia. That means we must really grasp the meaning of Marxism.

Briefly, the true Russian Communist believes, with the

* From *Why They Behave Like Russians*, by John Fischer, Harper & Brothers, New York, 1947.

devotion of religious fanaticism (as Mr. Fischer has pointed out in an earlier chapter), that capitalism, whose chief stronghold now is the United States, cannot escape from a cycle of booms and depressions, each worse than the last; that the capitalist ruling class will eventually turn in desperation to fascism, imperialism and aggressive warfare as the only way out.

Thus we may see that what appears to be ruthless conquest may be, at least in large measure, a defensive reaction based on fear of the vast military power and resources of the United States—a defensive reaction which may explain the Czech coup, the Communist-promoted strikes in France, the Communist support of guerilla warfare in Greece and the conflict associated with the Italian election. Marxist doctrine also explains the attempts by Russian delegates to sabotage the United Nations. For, from Russia's point of view, we are seeking to dominate that body and use it eventually as an instrument of our imperialism.

It may also be seen now that the expansion program engineered by Stalin and his associates is not the aggression of an Alexander, a Napoleon or a Hitler. The Nazi Fuehrer used powerful armies, even in Austria, where he had an organized mass following and several leaders in key positions.

Communist expansion is primarily political; it is a program of a group of nations relatively weak in respect to industrial resources, disturbed by the military power of a potential enemy, but not ready for a major war. While Russia has a large army, it is held in reserve to support measures essentially political. She may appear, from time to time, to be mobilizing military strength, but she does not really want her political advances to develop into a shooting war.

The Problem of Understanding Russia

To the extent that this analysis is sound, does it offer a hope for world peace? Can we match Russia's use of military force with sufficient force to stop further aggression? At the same time, can we match Marxist ideology with the ideology of democracy, exercising the patience and good will implicit in democracy in continued efforts of friendliness?

These questions bring us back to Mr. Fischer's conclusions. We must prove that our system can provide a better living and more freedom than the Communists offer. Soviet expansion may be stopped, and there may develop in a few years a fairly stable division of the world between the Communists and the Western democracies. Then the armed true may eventually—perhaps in another generation—be converted into peace.

In the meantime, can atomic energy, a leading cause of misunderstanding, be developed for use in constructive enterprises that may strengthen democracy here and abroad? David Lilienthal and other well informed individuals have answers to this question. Their views are now set forth.

· 11 ·

The Larger Implications of Atomic Energy

MOST OF US HAVE A FAIRLY ADEQUATE KNOWLEDGE OF the devastating power of the atom bomb, but little information regarding its constructive uses. Particularly limited is our understanding of the larger implications of atomic energy. The subject deserves the thoughtful attention of youths now in schools and colleges, and of older people who just now may have greater influence on public opinion and governmental policy.

Early predictions regarding the peacetime uses of atomic energy were soon found to be unjustified. Automobiles and private airplanes will not be operated on pink pills—at least for the present; houses will not be heated by spoonfuls of plutonium. The practical values of atomic energy will be limited for some years to come to two fields— (1) power plants and (2) research in industry and medicine. But even so, its uses, as we shall see, may later be revolutionary.

Work on the design of the first experimental atomic power plant was begun in 1947 at Oak Ridge, Tennessee. This plant will contain three components—a high temperature "atomic pile," a steam boiler, and a steam turbine

The Larger Implications of Atomic Energy 83

electric generator. The pile will take the place of the firebox of a conventional power plant, and it will make use of nuclear fuel instead of coal. Provision must be made for the safe disposal of prodigious quantities of by-products which are extremely dangerous because they are more radioactive than radium. It is hoped, however, that after several years, a productive power plant, based on the Oak Ridge experimental work, can be built at the Argonne National Laboratory, near Chicago.[1]

Atomic energy, it has been estimated, can compete with coal at $10 a ton. Some experts believe the cost differential between coal and nuclear fuel will be steadily reduced. Other experts feel that the $10 estimate is too low. But most of them agree that nuclear power has the potentiality of contributing a significant portion of our national power needs within the next few generations. The principal advantage of nuclear fuel is its lightness. One pound of U-235 or plutonium yields as much energy as 1,500 tons of coal. So the cost of transporting nuclear fuel is negligible. Thus, the first economically justifiable nuclear power installation will probably be built in a region remote from coal, petroleum and natural gas deposits.[1]

A word of caution is needed as to time scale. According to the general advisory committee of the United States Atomic Energy Commission, it will not be possible "under the most favorable circumstances to have any considerable portion of the present power supply of the world derived from nuclear fuel before the expiration of twenty years."[2]

In addition to power plants, atomic energy may be used in the future for the operation of locomotives and passenger liners. It can perhaps supply not only industrial power, but hot water or low-pressure steam for heating factories (which at present are costly)³ but this lies in the future.

One difficulty with atomic energy plants is that they release not only heat, but also horribly deadly radiation. The heat cannot be readily applied to generating steam, and it is expensive to stop the radiation. An atomic "pile" weighing as little as 100 *pounds* and generating 100,000,000 horsepower would require 100 *tons* of insulating material (lead, for instance).⁴*

In the field of research lie great possibilities—in discovering new ways of treating disease, particularly cancer, or (to quote a report by a panel of scientists) "in the increased understanding of biological systems or of the realities of the physical world, which will in turn open up new fields of human endeavor."⁶

* A group of process development engineers emphasizes the difference between two courses of action in the production of atomic energy.

First is the large-scale production of nuclear fuel for use in making bombs and for *possible* use in producing power. Plants for this purpose cost hundreds of millions of dollars; and there is little assurance, these engineers say, that there is enough material (uranium and thorium) in existence to assure the production of power for extensive long-term industrial use.

Second is the production of isotopes for basic research and for diagnosis of disease. Plants for this purpose cost as little as $10,000,000, and they can be operated independently of the production of bombs and power.

These engineers recommend the postponement of the first program, or course of action, until the world becomes more peaceful. Thus, international control of atomic energy might be more readily achieved with the cooperation of Russia; there would be no danger of seizure of atomic energy plants in the United States or Russia for military use; we could save a vast amount of money; and we would have far greater security.

Postpone for twenty years the costly program described first, these men propose, and we will lose no benefits which we now enjoy or can foresee in the near future. Leave the question of large-scale production to our children "who we pray will be better world citizens than we have been."

The Atlantic Region Working Committee of the Association of Scientists for Atomic Education submitted an "emergency" memorandum April 4, 1948, to the U.N. Atomic Energy Commission, urging a ban on large-scale atomic production for at least ten years.⁵

The Larger Implications of Atomic Energy

Radioactive isotopes* for research are produced by the Clinton National Laboratory at Oak Ridge, Tennessee. By August, 1947, this laboratory had made over 1,000 shipments of isotopes to 170 institutions and researchers in 92 locations in the United States. As a result, great progress has been achieved in the fields of medicine, biology, and industry.[7]

Approximately 100 varieties of radio isotopes have been produced at Oak Ridge. They are shipped in a specially constructed container. The isotopes are placed in a glass bottle inserted in an airtight stainless steel cylinder which in turn is placed inside a lead shield protected by a strong wooden box.[8] Radio isotopes are ticklish things to handle. Tiny specks of some of them can kill. The necessary skill is being rapidly acquired, however—the "know-how" as *Time* magazine says—"which is far more valuable than the perishable 'secret' of the bomb."[9]

At Memorial Hospital, New York, reports *Newsweek*, an intensive cancer research program providing for the use of radio isotopes is underway. At the National Institute of Health, Dr. R. E. Dyer has found that radioactive penicillin can be traced through the patient's body. This Institute is also using radioactive phosphorus to "tag bacteria in the study of tuberculosis bacilli." At Vanderbilt University, Dr. Paul F. Hahn has found radio manganese and radio gold "exceedingly useful" in the treatment of 108 patients with chronic leukemia, lymphoma, and Hodgkin's disease. Before isotopes were received, treatments averaged $92; now the cost is only about $6.[10]

Two more atomic laboratories are under construction—

* Variations of common elements with the same chemical properties as the stable elements but with different atomic weights and with the property of radioactivity.

one by the Atomic Energy Commission at Brookhaven, Long Island (the former site of Camp Upton), and the other by the University of Chicago at Chicago. At Brookhaven, nonmilitary uses of atomic energy will have a high priority. But until international agreements make it possible to concentrate entirely on peacetime uses, the great potential benefits of atomic energy may not be realized.[11] At the University of Chicago, the research program will be devoted exclusively to constructive uses. New buildings and equipment on which construction has been started will cost over $10,000,000.[12] Here will be three new subsidiary research institutes with such distinguished nuclear scientists as Enrico Fermi, Leo Szilard, Harold C. Urey, Edward Teller, and Thorfin R. Hogness. The research planned includes an attack on cancer and studies to discover whether steel can be made "at least ten times as strong as it actually is," which, according to atomic theory, is possible.[13]

The President of the United States announced September 2, 1947, at the fourth International Cancer Research Congress attended by scientists from 35 countries, that progress in this country in the production of radio isotopes "by the United States Atomic Energy Commission now permits limited distribution to qualified research workers in other countries of radio isotopes principally for medical and biological research."

This action of the United States was hailed by the scientists attending the Congress as a "momentous event" of great significance to mankind. Said Professor Shields Warren of the Harvard Medical School, "The United States is providing the world with the most important research tool since the discovery of the microscope."[14]

The Larger Implications of Atomic Energy

No human mind can predict what will emerge from discoveries still to be made in the field of atomic energy, asserts John W. Campbell, "any more than Volta could see in the future of his electric battery such things as radar contact with the moon, the great Hell Gate power station in New York, or the baffling mystery of a transformer. . . . We can err in our forecasts only on the side of conservatism."[15]

Now, in the pages which follow, we present the well considered views of David E. Lilienthal, chairman of the United States Atomic Energy Commission.

The dominant fact of our time [asserts Mr. Lilienthal] is the towering place of the machine, of applied science, in the lives of mankind. And the great issue of our time, with which the peoples of the whole world will be at grips day in and day out for the rest of our lives, is simply this: Are machines and science to be used to degrade man and destroy him, or to augment the dignity and nobility of humankind? How can men use and direct science and the machine so as to further the well-being of all men?

From this issue no one who lives today can escape. It reaches into the lives of every one of us, old and young, rich and poor; it must be faced by the clergyman, the professor, and the physician no less than by the businessman and public official. Standing always at the elbow of each delegate at Security Council meetings, always present and voting at the conference tables of this troubled world is this same issue: for whose benefit the machine? Cross the seas and the shadow of the question has preceded you—to the valley of China's Yangtze River, to the oil fields of Iran, the tractor-powered wheat farms of the Ukraine.

Geography and language differ, but the question is every where in essence the same.

Is science good, is the machine good—or is it evil?

To some people modern technology is plainly evil. To them the more gadgets, the more unpalatable is life. The more things we produce, the faster we can travel, the more complex machines we invent, the nearer—they assert—we move to the edge of a bottomless pit. They are homesick for the simpler life of the days before man produced so much and knew so much. They cry out against science and the machine and call them evil; but their voices are those of despair and defeat.

There are others of our contemporaries who have an almost opposite view of the machine. Exuberant and uncritical, they worship the machine. Efficiency is their god and the managerial elite are their high priests. They broadcast radio programs full of the romance of gadgets and fill the slick-paper magazines with odes to a chromium bathtub. Technology, they seem so say, is above good and evil. If the spirit of man balks, if being human increases the cost of production—well, then man must be redesigned to fit the assembly line, not the assembly line revised for man. The supertechnologists of the world are quite prepared to recreate man in the image of the machine.

I venture to say that neither of these views will appeal to most of us. The machine is neither good nor evil in itself. It is good only when man uses it for good. It is evil only if he puts it to evil uses.

The machine can, of course, be so used as to degrade and enslave man. It can be used to exhaust the land and the dignity of those who live on the land. It can poison the air, foul the streams, devastate the forests, and thereby

The Larger Implications of Atomic Energy

doom men and women and children to the degradation of great poverty. On the other hand, it can nourish the spirit of men, eliminate filth and congestion and disease, strengthen the soil and conserve the forests, humanize man's environment. The machine can be so used as to make men free as they have never been free before.

We have a choice—that is the shining fact of our time. If we are but wise enough and determined, if we have the knowledge and the facts on which to base our choices, we can control the machine and make it serve for good. We need not be carried irresistibly into the abyss by forces beyond our control.

We have a choice—whether the machine or man comes first. But that choice will not be exercised on a single occasion, surrounded by spectacle and drama. We must move from decision to decision, from issue to issue, and we will be in the midst of the struggle for the rest of our days.

In such a setting there is one generalization that I believe is basic. It is this: We cannot master the machine in the interest of the human spirit unless we have a faith in people. The rock upon which all these efforts rest must be a deep and abiding faith in human beings, a faith in the supreme worth of life. Technology must have a moral, an ethical purpose; that purpose and the methods pursued in carrying it out determine whether technology furthers human well-being or threatens it.

Great schemes of development are now being unfolded for China and India and Siberia and Latin America and the Near East—regions technology and science have hardly touched. And we are asked: But does it really matter what means are adopted to bring science to the millions in these undeveloped lands? Does it really matter that the purpose of such development may not be human welfare? Does it matter really whether the people concerned have information with which to make their own choices, whether they have a genuine voice in this new destiny that science and the machine can bring? In brief, does it matter whether the purpose and the means are moral and ethical? Matter? Indeed it matters! Nothing matters more profoundly.

The very act of faith in the essential goodness of men furthers that goodness, just as disbelief in men will of itself make men less worthy of faith. This is not mere rhetoric; these are the ultimate realities. It is by putting to practical use these realities that science and technology, by our deliberate choice, can bring a new era of fraternity among the peoples of the earth.

All of the foregoing has its immediate and critical application in the field of atomic energy. Our task, in practical terms, is to turn the awful strength of atomic power away from destruction and toward a new era of peace and well-being.

There is nothing more dewy-eyed, naive, and ignorant of the realities of human nature than to urge (seriously and sincerely) that the world rest its hopes for eliminating the atomic bomb entirely upon making possession and use of the bomb illegal—"outlawing" it is the common phrase—and providing machinery for drastic punishment of violations.

Suppressive and police-like activities, taken alone and relied upon exclusively, are no foundation for peace—something more affirmative is required. It seems clear to me that we must, if we would survive, seize upon, emphasize, and make use of the creative opportunities. It seems

The Larger Implications of Atomic Energy

clear that it is these very constructive possibilities in respect to atomic energy that make this such a great and unique opportunity for the building and maintenance of peace.

My associates and I were explicit about this in our report.* We do not share the feeling that to build the edifice of peace upon such a foundation is unrealistic. In describing the criteria we laid down to guide us in devising a plan for international control of atomic energy we said: "To be genuinely effective for security, the plan must be one that is not wholly negative, suppressive, and police-like. We are not dealing simply with a military or scientific problem but with a problem in statecraft and the ways of the human spirit. Therefore, the plan must be one that will tend to develop the beneficial possibilities of atomic energy and encourage the growth of fundamental knowledge, stirring the constructive and imaginative impulses of men rather than merely concentrating on the defensive and negative. It should, in short, be a plan that looks to the promise of man's future well-being as well as to his security."

We cannot think of atomic power as apart from those who direct its course and use, and define the purposes for which it is put. That is a towering responsibility which may safely be entrusted only to those who have a deep faith in the human spirit, to whom the interests of human beings come first.

We do have a choice. On the one hand it is clear that science in evil hands can make us slaves—well-fed perhaps, but more pathetic for that fact. On the other hand it is

---

* A report on the international control of atomic energy, by a five-man Board of Consultants, made available by the State Department, March 16, 1946.

plain that men can use science to further human freedom and the development of human personality.*

If the world is to survive the emergency of the next decade or two, it will be necessary that young people, as well as old, have an understanding of international problems, of Russia, of war and human nature, and of the larger implications of atomic energy. For soon our older youth, now in high school and college, may be occupying strategic positions. The importance of their education in these fields should surely be considered.

---

* Condensed from "Science and Man's Fate," by David E. Lilienthal, *The Nation*, July 13, 1946. With the permission of the author and editor.

· 12 ·

Training and Understanding for Youth

"THIS LETTER MAY NEVER BE DELIVERED. IT WILL GO TO Corregidor and there await transportation." [Thus wrote an American soldier to "Mother and Dad" in the last few hours before the fall of Bataan.]

"I am proud [he continued] to be a part of the fight that is being made here and would not, even if it were possible, leave here until it is over. . . .

"I have seen some horrible things happen, but I have also seen wonderful acts of courage, sacrifice and loyalty. At last I have found that for which I have searched all my life—a cause and a job in which I can lose myself completely and to which I can give every ounce of my strength and my mind. . . .

"During the last two months I have been a part of one of the most unselfish cooperative efforts that have ever been made by any group of individuals. Mistakes have been made, but that has nothing to do with the manner in which my comrades on Bataan, both Filipino and American, have reacted to their trial by fire. If the same spirit were devoted to world betterment in time of peace, what a good world we would have. . . ."*

* From the Washington *Daily News*, and reprinted in the *Reader's Digest*, September, 1942.

Veterans are back from the war and are crowding the colleges of the country. Many, of course, want technical training; others may be seeking, more or less consciously, an understanding of the forces which lead to frustration, injustice, and war. Some of them, perhaps only a few, may be looking for a "cause and a job" in which they can lose themselves completely and to which they can give all their strength and all their minds. Ernie Pyle, in writing of the audacity of the European invasion and the power and strength of our armies, said he had heard soldiers say a thousand times, "If only we could have created all this energy for something good."[1]

Can veterans and other youth find, in times of peace, opportunity to fight for "something good," opportunity for "unselfish, cooperative efforts"? Can they find "a cause and a job" in the building of a new and peaceful world?

Here before us, today, is a cause. Consider the menace of communism, the hunger and chaos of Europe and Asia, the present inadequacies of our own democracy, the armament race now underway, the atom bomb hanging over us like the Sword of Damocles, and the terribly difficult and laborious task of building a friendly world—a world of freedom and justice for all peoples—with a strong international organization to maintain the peace. Here is a cause more challenging and more demanding than the cause of freedom for which our forefathers fought, a cause requiring intellectual and moral resources as great as those possessed by the men who wrote the Constitution and brought together the quarreling colonies into a strong union.

Jobs, too, can be found. In various voluntary organizations working for a united world,* whole-time positions are

* See Appendix B, p. 178.

Training and Understanding for Youth

occasionally available for well-trained persons. There are also innumerable part-time openings for informed leaders—an essential of democracy—in businessmen's clubs, women's organizations, educational agencies, and political groups. A government position may well be the goal for some individuals.

At Washington there are opportunities for a life work in the State Department, especially in the consular and diplomatic services. The State Department is desirous of replacing some of the restless, wealthy dilettantes with able and ambitious, but not necessarily rich, new men. Senators and representatives also are needed who have been thoroughly trained in foreign relations. Experts on Russia will be invaluable for years to come as members of the Senate and House Committees on Foreign Affairs. Trained economists, food experts, anthropologists, international lawyers, as well as specialists in monetary problems, the tariff, and other fields are needed in various congressional committees.[2]

In the Secretariat of the United Nations are unique opportunities. At present the Secretariat consists of about 2,700 persons. The Secretary-General is the chief administrative officer of all bodies in the United Nations except the Court. He and his various colleagues are persons of considerable importance. While many applications are being received, positions are available to outstanding candidates.[3]

The present situation in international affairs demands statesmen with the idealism of Woodrow Wilson and the political sagacity of Georges Clemenceau, Franklin Delano Roosevelt, and Winston Churchill. Politics needs young men and women. In England a large proportion of the ablest students in Oxford, Cambridge, and the other universities go directly into the service of the government. In our own history, Jefferson, a graduate of William and Mary College, became a member of the Virginia House of Burgesses at the age of twenty-six. James Madison went to Princeton, and at twenty-five was elected a delegate to the Revolutionary Convention of Virginia. Theodore Roosevelt, Robert M. LaFollette, Jr., and various young men in modern times have followed their examples.

Yes, jobs will be available to the youth—provided we can outlive the crisis of the next five to ten years. But he must meet the requirements; he must be willing to do something more difficult than to give his life to a great cause—he must be willing to endure adversity and face discouragement without the dramatic experiences which often ennoble the hero on the field of battle. (In military warfare, it may be remembered, those who die quickly in the heat of battle are in the minority; the large majority have to endure the grime, filth, hardships, and monotony of war without its thrilling experiences.) He must also be willing to go through a period of training—preparation sometimes requiring many years of study.

The training necessary consists of academic study and various extracurricular activities both in and out of school. The important academic subjects are in the field of the social sciences—economics, sociology, government, psychology, anthropology, history, and similar subjects.

The causes of war should be studied from various points of view, and the contemporary problems of Europe and Asia thoroughly understood. Russia, the master problem, requires special attention. Finally, there is atomic energy, to be approached both as a problem in physics and chemistry and as a problem in the social sciences. Today the world

Training and Understanding for Youth 97

needs social scientists more than it needs men skilled in the natural sciences. And, as explained in an earlier chapter, the making of a social scientist, no less than the making of a chemist or an electrical engineer, demands intensive application to an extensive and rapidly growing body of subject matter.

The colleges of the country are rapidly adding to their programs of study, courses in such subjects as the Rise and Fall of Fascism, International Relations, Soviet Russia, the United States and Foreign Affairs, World Economic and Social Problems, Far Eastern International Relations, and Internationalism. A "workshop" on international affairs, in New York in 1947, gave special attention to China; assistance was provided by a staff of Chinese professors from the China Institute of New York City.[4]

Many colleges offer organized "sequences" of study in various phases of international relations. These include Stanford University, the University of Denver, George Washington University, Carleton College, Colgate University, and Middlebury College.[5]

A "World School" is to be established by the Riverdale Country School in New York, with initial donations of more than a quarter million dollars. A 28-acre tract has been purchased, and a fund of 5 million dollars is being sought. The plans provide for 600 teen-age students—200 from Latin America, Europe, and Asia. Attaches of the United Nations will be invited to recommend studies in the problems of their various countries. Visits will be made to the United Nations sessions.[6]

Seniors at Dartmouth College are required to take a "Great Issues" course. It deals with the control of atomic energy, the development of adequate world government, and modern economic problems. Monday evenings guest lecturers present their views on special subjects, and Tuesday mornings these views are discussed by the students. The course is taken by 600 seniors divided into twenty or twenty-five groups. The Carnegie Corporation, recognizing the significance of the course, has agreed to provide substantial financial aid for the first three years.[7]

One way to gain understanding of the international situation is to get into action. The organization in school or college of a student "United Nations," of which several sessions may be held, affords such an opportunity. The American Association for the United Nations will assist any group wishing to organize one.* This Association conducted a prize contest recently in which 1,925 high schools participated. The questions pertained to the functions of the United Nations and its subsidiary and affiliated bodies. Beatrice Hauser of Sheboygan, Wisconsin, won the first prize of $500. Nearly fifty other prizes have been offered by state and local branches of the Association.[8]

Student Federalists, a division of United World Federalists, presents a fine opportunity to students, both in college and high school, to work for the kind of world government this organization believes will assure a united, peaceful world. In institutions where chapters are already established, students can find opportunities for useful activities. Where there are no chapters in existence they may be organized with the help of the national office. There are also a considerable number of local chapters of the Youth Council on the Atomic Crisis.*

Institutes of International Relations under the auspices

* See Appendix B, p. 178 for a list of agencies working for world peace.

Training and Understanding for Youth

of the American Friends Service Committee were held during the summer of 1947 in thirteen locations for college students and in eight locations for high school students. Each of the two groups covered territory from California to Massachusetts.[9] In 1948 Institutes were held in approximately the same locations.

Twenty-four students from eleven eastern colleges reported for work at the United Nations Headquarters on July 10, 1947, as "internes" in international organization, under a new program arranged by the United Nations Training Division. These students were unpaid and worked from 9:30 A.M. to 6:00 P.M. for eight weeks. Three days each week they attended lectures by top Secretariat officials. On another day a two-hour seminar was held. Similar programs were developed for 1948—one in the summer especially for students outside the United States.[10]

A particularly promising project in the summer of 1947 provided for the sending of American students abroad, and for the reception of foreign students by the educational institutions of this country. These students made a total of about 3,500 trips, one way or the other; a minority were foreign students. This exchange was arranged by our State Department. Two government transports were made available for four round trips each to Europe, but the passengers paid their own way. One contingent of 800 students, teachers, and others sailed on June 30th. It contained, among other groups, some Yale University graduate students who were to study British educational methods. An American Youth Hostel group was to spend about six weeks working with Norwegians.[11] Before the ship sailed a *bon voyage* ceremony was held, with Kenneth Holland, of the State Department, as one of the speakers. "It is appropriate," he told the group, "that you should be sailing on a ship which once carried thousands of troops to distant battles. You are fighting another phase of the same war your brothers fought by force of arms. . . . We are still fighting a war against ignorance, hate and prejudice." The Norwegian Ambassador, Wilhelm Muth de Morgenstierne, said he knew of nothing that in the long run carries more hope for human brotherhood and peace than personal acquaintance and understanding among young people of different nations.[12]

Under the Fulbright Act, an estimated $140,000,000 will be available during the next twenty years for international educational exchange. A maximum of $1,000,000 per year may be expended in each foreign country which has concluded an agreement with the United States. The program includes the exchange of students and educators. Unfortunately, because the grants must be made in the currencies of the foreign countries involved, their citizens coming to the United States will be able to obtain assistance covering their travel costs only. Americans going abroad will receive grants including tuition, maintenance, and incidental expenses as well as travel. Thus, while thousands of Americans are able to study abroad, relatively few foreign citizens can come to this country. Others must postpone their foreign training or turn to countries other than the United States. It is known that the Soviet Union is providing a sizeable number of scholarships for students in nearby countries.

Students from Latin American countries can come to the United States more freely, a small number of them assisted by the Department of State; and United States students may go to Latin America. The Smith-Mundt Act of January 1948 extends this type of reciprocal program to other areas, including the devastated countries in Europe. This act supplements the Fulbright Act admirably and makes the opportunities for educational exchange world-wide.[13]

To the superior student with imagination and initiative, the present international situation constitutes a real challenge. There are great tasks ahead for the youth who has acquired the necessary training, has tried himself out, has weighed the cost, and is ready for high adventure and bold action.

The conquest of war is not a negative goal sought by soft, faint-hearted, and tired persons who want a dull peace. Man's warfare against war and the great task of building a new world require not only the physical courage so often demonstrated at Guadalcanal and on many other fields of battle, but also the open mindedness, the energy, and the moral courage and idealism of youth. Adults often resist change, if for no other reason than because it *is* change. It is to the youth of the world that we must now turn—the youth who has thoroughly prepared himself in school and in practical close-up contacts with the current affairs of our critical postwar world.

PART III

Which describes various ways of building a united, peaceful world

W<small>AR IS MORE TERRIBLE THAN ALL THE WORDS OF MEN</small> can say: more terrible than a man's mind can comprehend. It is the groans and the pain of the wounded, and the expressions on their faces. It is the filth and itching and hunger; the endless body discomfort: the fatigue so deep that to die would be good. It is battle, which is confusion, fear, hate, death, misery and much more. It is "battle fatigue," a nice name for having taken more than the brain and heart can stand. The reality—not the words. Remember!

And when it is in your mind so strongly that you can never forget: then seek how you can best keep peace. Work at this hard with every tool of thought and love you have. Do not rest until you can say to every man who ever *died for man's happiness:* "You did not die in vain."

—Condensed from a letter which an American soldier in Germany wrote to his son.*

* By Walter J. Slatoff. Used with his permission and that of the New York *Times*.[2]

· 13 ·

The United Nations

ONE OF THE GREATEST EVENTS IN HISTORY HAPPENED AT San Francisco on January 26, 1945, when the newborn charter of the United Nations was signed by the representatives of fifty nations. [Here the American Association for the United Nations describes the new organization.] Other events had led up to that day—the adoption of the Atlantic Charter in August of 1941, and back of that the first League of Nations, founded in January, 1920, and still earlier the Hague Conferences, and, spreading over many centuries, the plans and dreams of philosophers and statesmen.

Although the United States was not a member of the League of Nations, we are now part of the United Nations, and, with our great power and wealth, an important part. We are no longer on the outside looking in. We carry a responsibility for its success or failure. This being true, we should surely become well-informed regarding the new organization.

The broad purposes of the United Nations are (1) to maintain international peace and security by preventing aggression and settling disputes peacefully; (2) to develop friendly relations among nations based on respect for equal rights and self-determination of peoples; and (3) to achieve international cooperation in the study of economic, social, cultural, and humanitarian problems and in the promotion of human rights and fundamental freedoms.

The United Nations is composed of fifty-seven member nations. Its "organs" are (1) a General Assembly of all nations, each with one vote, responsible for the policies of the world organization; (2) a Security Council of eleven, in continuous session, to take action—diplomatic, economic, or military—to prevent or stop wars; (3) an Economic and Social Council of eighteen members to promote the well-being of peoples and respect for human rights and fundamental freedoms; (4) a Trusteeship Council to supervise the administration of dependent peoples under the world organization, and to help them toward self-government or independence; (5) an International Court of Justice for the settlement of legal disputes; (6) a Secretariat to do the day-by-day work, with a Secretary-General as its chief officer.

The General Assembly consists of all member nations, each of which has one vote. It meets at least once a year. Each member nation may send one to five representatives. It is a great public forum where its members may discuss any matter within the scope of the Charter. It makes recommendations to the three councils and receives reports from these councils as well as from all other branches of the U. N. organization.*

The Security Council has "primary responsibility for

* The Interim Committee of the General Assembly, which meets continuously, was set up in November, 1947. It is called the "Little Assembly." It is expected to strengthen the parent body.—Editor.

The United Nations 107

the maintenance of peace and security." Its eleven members are: China, France, the Union of Soviet Socialist Republics, the United Kingdom, and the United States (the Big Five), which are permanent members, and six nonpermanent members elected by the General Assembly for two-year terms. The Council functions continuously with each member nation represented at headquarters at all times, and it is busily engaged with disputes brought before it by members.

Broadly speaking, the Security Council has two functions, (1) to see that any disputes between nations are settled peacefully, and (2) if peaceful settlement fails, to prevent or stop any act of aggression. If a dispute is not settled by this means, the Council may call upon members of the United Nations to stop trade with the offending nation and break off communications and diplomatic relations. Thus the offender can be completely blockaded and cut off. If the above means are used and prove ineffective, the Council may use land, sea, or air forces in demonstrations, blockades, or "other operations." For this purpose member nations must be ready to provide armed forces for such operations. The United Nations now has no police force of its own.

It will thus be seen that the Security Council has great power and great responsibility. Routine matters may be decided by a vote of any seven of the eleven member nations. In matters involving action, or investigation which might lead to action, the majority of seven must include the five permanent members even if one of them is concerned in the case. The net effect of this provision is that *no enforcement action can be required from or against one of the Big Five.*

The Economic and Social Council of eighteen members is responsible for an important objective of the United Nations—the promotion of "higher standards of living, full employment, conditions of economic and social progress . . . international cultural and educational cooperation . . . universal respect for . . . human rights and fundamental freedoms for all without distinction as to sex, race, language, or religion."

The Trusteeship Council has general direction of territories which may be turned over voluntarily to the United Nations—(1) those held under League of Nations mandates, (2) territories taken from Axis powers, and (3) dependent territories.

The International Court of Justice consists of fifteen judges elected by the General Assembly and the Security Council. Members are not obligated to submit disputes to the Court (unless by treaty), but if they do submit a case, they promise to obey the Court's decisions. Should any nation fail to do so, an appeal can be made to the Security Council which may take any action it sees fit.*

Among the principal Commissions of the United Nations are (1) the Atomic Energy Commission, (2) the Commission on Conventional Armaments, and (3) the Human Rights Commission.

Of these, the Atomic Energy Commission has been most conspicuous in the news. This Commission recommended in December, 1946, the creation of an International Control Agency, effective inspection and assurance of prompt punishment for violators unimpeded

* The fore-going description of the United Nations is based on the booklet, "We, the People. . . . ," published by the American Association for the United Nations.

The United Nations

by the action of any one government. The recommendation, however, was not approved. Unfortunately, after many meetings of the Commission no plan for the control of atomic energy has yet been advanced which has been acceptable to both the Russian and United States delegates.

The International Children's Emergency Fund (ICEF), is an integral part of the United Nations, and has a technical staff which assists governments, at their request, to formulate plans in behalf of the welfare of their children. There were 30,000,000 children and nursing and pregnant mothers in dire need in Europe in February, 1948, and at least that number in the Far East. By "dire need" ICEF means "actually starving, or on the verge of starving." Supplies had been obtained from many countries, early that year, for a feeding program. These included enough dried, whole skim milk to make 140,000,000 quarts, most of it from the United States; cod liver oil in quantity from Canada and other countries; lard, margarine, butter and meat from Canada, Australia and the United States. While the amounts on paper seem large, actual supplies were far short of needs. A drive to "Give One Day," early in February, 1948, had as its goal $60,000,000, of which $21,000,000 was to be earmarked for the ICEF, the remainder to be allocated among 26 voluntary relief agencies.[1]

Is the United Nations a success? After three years, how effectively is it functioning?

One answer is that it has not brought peace to the world. Fighting was in progress, during 1947, in China, Greece, Palestine, and Java. About 17,500,000 men, as has already been noted, are now under arms in forty countries which are spending approximately $30,000,000,000 a year on armaments, plus huge additional sums for military research. Labor, wealth, and scientific knowledge, withheld from the mainstream of productive activity, constitute one of the heaviest drags on the world's slow progress toward a more peaceful future.

The Commission to Study the Organization of Peace, a research affiliate of the American Association for the United Nations, issued a report in June, 1947, which contained an appeal for the support of public opinion, and a series of recommendations for United Nations action on security and disarmament.

Despite all the obvious difficulties [asserts the Commission], the United Nations can do more than it has done to organize the peace of the world. It has actually made more of a start than most impatient people realize, but it can be speeded up. Action on disarmament and security is possible now, and it can be successful if world public opinion gets behind it. Peoples everywhere want peace and the assurance of a secure future. This popular will is the strongest power on earth. It must make its way to the dominant place at the conference tables of the nations.

One serious weakness of the UN has been its inability in many cases to arrive at prompt decisions. Unwarranted use of the veto power has been in part responsible for this weakness. *The Commission does not at this time propose to amend the Charter's provisions on the veto,* but proposes methods whereby the UN's ability to reach decisions will be strengthened without eliminating the veto principle.

We appeal [continues the Commission] for a world public opinion in support of the United Nations. We appeal particularly for an American public opinion which will support our government in taking bolder initiatives in the interests of peace, and we are certain that such leadership will have the overwhelming support of the American people.

The United States mission to the United Nations should continue with perseverance, patience, and courage to insist upon an approach to the total problem of security and disarmament.*

A fair judgment of the success of the United Nations cannot be formed without considering the achievements of some of the "specialized agencies" connected with it. There are at present eleven of these agencies which are affiliated with the United Nations through the Economic and Social Council. Four of them were in existence before the United Nations Charter came into force. They all enjoy considerable independence. On the other hand, their connection with the United Nations is quite real and the Charter authorizes the Economic and Social Council to "take appropriate steps to obtain regular reports" from all of them.

Prominent among the specialized agencies is the "United Nations Educational, Scientific and Cultural Organization" (UNESCO). It is made up of 40 member states; the budget adopted for 1948 was about $8,000,000—about 1/80, remarks the UNESCO, of one day's cost of World War II. "Since wars begin in the minds of men," declares the UNESCO constitution, "it is in the minds of men that the defenses of peace must be constructed." The purpose of the organization is "to contribute to peace and security by

* From a Summary of the Report of the Commission to Study the Organization of Peace, *Newsweek*, June 16, 1947, pp. 35, 36.

promoting collaboration among the nations through education, science and culture."

The program adopted for 1948 covers six major fields—reconstruction, communication, education, intercultural interchange, human and social relations and natural sciences. Within these fields are nearly one hundred specific projects. UNESCO already has purchased $70,000 worth of war surplus and new material, especially scientific apparatus, which has been distributed to technical schools in devastated countries. It sponsored, in the summer of 1947, four international youth camps in Belgium, Poland, Czechoslovakia and France. Young volunteers from Europe and America worked during the day on the reconstruction of war-damaged schools and homes; the evenings were reserved for reading, discussions and excursions.[2]

Another specialized agency in the Food and Agriculture Organization (FAO), created to improve the nutrition of the peoples of all countries, and to increase the efficiency of farming, forestry and fisheries. Missions of technicians have been sent to Greece, Poland, Venezuela and Siam (whose governments requested them) to aid in increasing nutrition and agricultural production. Special schools have been conducted for agricultural technicians in Europe.[3]

The International Refugee Organization (IRO) had not come into being early in 1948 by the ratification of its constitution. But a Preparatory Commission had undertaken certain functions in respect to 640,000 refugees and displaced persons in Germany, Austria and Italy. Its main tasks include the care of persons now in camps or assembly centers not yet returned to their homes, and their repatriation or resettlement as soon as possible. For the year which ended June 30, 1948, its budget was $119,000,000.[4] The

The United Nations 113

Preparatory Commission took over part of the work of the United Nations Relief and Rehabilitation Administration (UNRRA).*

The World Health Organization (WHO), became a specialized agency when, in April, 1948, 26 UN countries had ratified its charter. Previously, an interim commission had adopted a $6,000,000 budget, providing for fellowships, teaching equipment and medical supplies to meet post-war health problems, world-wide campaigns against malaria, tuberculosis and venereal diseases with a top-priority program for mother and child welfare. This agency's greatest accomplishment in 1947 was its record-breaking fight against the cholera epidemic in Egypt. In China a WHO mission of 28 individuals has been working in cooperation with Chinese officials and agencies in an effort to control cholera and plague. Special attention has been given to the teaching of Chinese medical personnel. In Greece a mission has been concentrating on malaria and tuberculosis control. A dozen war-ravaged countries have sent approximately two hundred Fellows, under a $1,500,000 grant, to study at world centers of medical knowledge.[6]

Other specialized agencies include the following: The International Bank for Reconstruction and Development, the International Monetary Fund (IMF), the International

* At this place it may be appropriate to mention some of the achievements of UNRRA in the public health field. Europe's and Asia's amazing escape from pestilence during the war was partly due to this agency. It shipped 3 million pounds of DDT, 5,000,000 units of penicillin, 900,000 pounds of sulfa drugs, 13 million cubic centimeters of diphtheria toxoid, 8,000,000 cubic centimeters of diphtheria anti-toxin. By the end of 1946, typhoid, which had caused Europe's most serious postwar epidemic, was under control, diphtheria had been greatly reduced, typhus was rare, smallpox and plague had virtually been wiped out. While the United Nations cannot be credited with this fine work, such a record again shows what can be done by organized international effort.[5]

Trade Organization (ITO), the International Labor Organization (ILO), the International Civil Aviation Organization (ICAO), the Universal Postal Union (UPU), and the International Telecommunications Union (ITU).

Many well-informed and public-spirited citizens have faith in the United Nations. They believe that, through patience and perseverance and fair treatment, the United States and other democratic nations can get along with Russia—in the United Nations and its various subsidiary agencies. They fear that an attempt now to replace the United Nations by an international organization with greater power than that exercised by the United Nations would fail, and produce a dangerous reaction. Other equally well-informed and public-spirited persons believe that a world government with power to enact and enforce law is now necessary.

We turn, therefore, to a consideration of various proposals for world government.

· 14 ·

Proposals for World Government

IN 1787, A LOW-WATER MARK HAD BEEN REACHED IN THE history of the American colonies [Edwin Muller relates the story]. The Revolutionary War had been won, the Continental Congress had met, and the Articles of Confederation had been ratified in 1781. But still there was no peace.

"There are combustibles in every state which a spark might set fire to," wrote Washington. "I feel infinitely more than I can express for the disorders which have arisen."

For Shays's Rebellion was only one of the many "disorders." In western Massachusetts, in Vermont, and elsewhere in New England there were armed clashes. In New York the militia of Dutchess and Columbia counties were called out. There had even been the beginnings of actual warfare between the states.

Foreign observers commented on our affairs with complacent I-told-you-so's. For example, the Dean of Gloucester: "As to the future grandeur of America, it is one of the idlest and most visionary notions that ever was conceived. The mutual antipathies and clashing interests of Americans, their difference of governments, habitudes and manners, indicate that they will have no center of union and no common interest. A disunited people to the end of time, suspicious and distrustful of each other, they will be divided and subdivided into little commonwealths or principalities according to natural boundaries."

This wasn't really a nation. It was merely an alliance of thirteen independent republics straggled out on a long seacoast. The alliance was held together shakily by the Articles of Confederation. In effect the Articles comprised a treaty by which the thirteen states agreed to act together —as the United Nations agree today. The alliance had been able to win a war. But, as usually happens, when the war was over it began to disintegrate. Its members followed their separate interests.

The only machinery for acting together was Congress. And Congress was little more than a council of ambassadors. There was no central executive power. There was a President, the president of Congress, but he had no more authority than any other member. Congress was weak because it had no effective way of enforcing its laws. The central government could neither raise money, maintain an army and navy, nor establish trade or other relationships between the states.

There were some Americans who saw the remedy. Washington was one. He insisted that the only hope was a real union under a single federal government. But the average American wasn't for it—not yet. The states were not willing to surrender any part of their sovereignty to a "superstate"—a word much in vogue. There was a heavy, inert mass of resistance to the making of a nation. To overcome it required a crusade, as daring and forceful as that which had brought about the Revolution.

Proposals for World Government

Alexander Hamilton, with Washington, Madison, and others, proposed that Congress give its sanction to a convention of delegates from all the states to make certain revisions in the existing Articles of Confederation. The legislatures received the proposal without enthusiasm.

Finally, fifty-five delegates assembled in Philadelphia in May, 1787. These delegates moved from step to step, sometimes a little shocked at the novelty of what they were doing. By fitting together their different concepts they worked steadily toward their goal—a national government which should be strong and centralized, yet in which the states should not be submerged. The work was done at last. The Constitution was submitted to us, the people. On the whole we didn't like the looks of it. There was a clear majority in the country against its adoption.

The common man felt that something had been put over on him. This went too far. He saw tyranny ahead. Tyranny of Congress, which could control elections. Especially tyranny of the President. The Constitution was called a conspiracy of the well-born against the common people. Through all the thirteen states the contest developed. There were outbreaks of violence. There was a pitched battle in which swords and bayonets were used. These were the dark days of 1787.

But slowly the tide turned in favor of the Constitution. It turned because the average man came to see the alternatives. On the one hand increasing chaos. On the other, a strong central government. The popular will was expressed in state conventions. Delaware was the first to ratify on December 6, 1787. Rhode Island was last, on May 29, 1790.

So we the people took the Constitution—a little uncertain whether we had a bargain or not. Now we know.*

The difficulties encountered by the United Nations have led many leaders of public opinion to believe that we now have in the United Nations Charter a document comparable to the Articles of Confederation. And just as the Articles of Confederation in our early history were not adequate for the maintenance of the peace and the development of colonial prosperity, so now the United Nations Charter cannot be relied upon as an instrument to bring about a united peaceful world. Various influential individuals and national agencies, therefore, propose the creation of a Federal World Government. Two ways have been suggested for the attainment of this end—(1) a revision of the Charter of the United Nations, and (2) the setting up of a world government independent of the United Nations.

Proposals for Charter Revision

Dr. Albert Einstein would like a world government established "through the strengthening of the United Nations." He is "in favor of inviting the Russians to join a world government." If they are unwilling to do so, he would proceed without them. But he would make it "utterly clear that the new regime is not a combination of power against Russia." If a world government is set up without Russia, it must be clear to Russia and the other nations that "no penalty is incurred or implied because a nation declines to join." If the Russians do not join at the beginning,

* Condensed from "Our Postwar Problems of 1787," by Edwin Muller, *Reader's Digest*, February, 1945, with the permission of the author and publisher.

Proposals for World Government

"they must be sure of a welcome when they do decide to join."[1]

One proposal for the creation of world government by revising the Charter is sponsored by the United World Federalists, Inc.* This organization believes that peace is not merely the absence of war, but the presence of justice, of law, of order—in short, of government and the institutions of government; that world peace can be created and maintained only under a world federal government, universal and strong enough to prevent armed conflict between nations, and having direct jurisdiction over the individual in those matters within its authority. "Therefore," the UWF by-laws explain, "while endorsing the efforts of the United Nations to bring about a world community favorable to peace, we will work to create a world federal government with authority to enact, interpret and enforce world law adequate to maintain world peace:

"(1) by making use of the amendment processes of the United Nations to transform it into such a world federal government;

"(2) by participating in world constituent assemblies, whether of private individuals, parliamentary or other groups seeking to produce draft constitutions for consideration and possible adoption by the United Nations or by national governments in accordance with their respective constitutional processes;

"(3) by pursuing any other reasonable and lawful means to achieve world federation."

High officials of Canada, Australia, New Zealand, the Netherlands, Great Britain, China, the Philippines, France,

* A list of agencies, with addresses, working for world peace are included in Appendix B, pp. 178 ff.

and Cuba have expressed a willingness for their various nations to abrogate at least a part of their national sovereignty in order to establish a stronger UN, or to transform it into a world government.[2]

A second proposal for revising the Charter of the United Nations provides "a specific blueprint for immediate action." It is called the "Quota Force Plan," and it is sponsored by the Citizens Committee for United Nations Reform, Inc.

The committee recommends a series of thirteen measures, embodying the Quota Force Plan as their main feature. In general, the plan provides for the reorganization of the Security Council of the United Nations and the abolition of the veto right of major states "in matters specifically pertaining to aggression and preparation for aggression."

The Council would fix yearly the total quantity of heavy armaments to be produced in the world and would maintain under its direct control an International Contingent which would be a highly paid professional armed force of volunteers recruited exclusively from citizens of smaller member-states.[3]

The Quota Force Plan was adopted by the American Legion on July 23, 1947, after a year's study. The recommendations in the plan have been sent to each member of Congress.[4]

PROPOSALS FOR A NEW WORLD GOVERNMENT

Proposals for a world government are advocated by several groups and many individuals. Dr. Harold C. Urey, atomic scientist of the University of Chicago, here sets

Proposals for World Government

forth reasons for by-passing the United Nations and establishing a new world government.

There is no real solution to the atomic bomb problem [asserts Dr. Urey] except the complete solution to the problem of war. It is my conclusion that only world government can prevent war in the future.

What is world government? The United Nations does not represent a government. It is analogous to the organization of this country under the Articles of Confederation between 1781 and 1789. It has no sovereignty. There is no way of making laws or of enforcing them. It has no way of supporting itself except by passing the cup to 57 nations.

A world government must have some properly constituted organization that can make laws, an executive body that can enforce them, and courts that can make decisions when laws are violated. It must have taxation to support itself, for he who pays the piper calls the tune. Moreover, a government must have direct access to its citizens—i.e., it must make laws for its citizens, and enforce them directly upon its citizens. This relationship was discussed in this country in 1787, and its effectiveness has been demonstrated by the success of our own federal government, as well as of others, since that time.

I believe that a world government including Russia and her satellites, as well as the western democracies, is not possible. Temporarily, at least, an all-embracing world government cannot be secured.

It is, therefore, necessary to consider limited alternatives. As matters stand, the world is divided into two groups, the U.S.S.R. on the one hand and the U.S.A. on the other, together with a considerable number of countries in various parts of the world, but particularly in Western Europe, that are in the tragic position of lying between these two centers of power.

My proposal is that we set up a federal union of as many countries of the world as possible. I believe that a substantial area of agreement exists between the Western democracies. Their governmental structures, while disagreeing in details, are in an overwhelming degree similar to each other. These are all in marked contrast to what we find in the totalitarian countries.

What would be the effects of such a limited world government, in which the word "limited" must be applied to the word "world" as well as to the word "government"? In the first place, it would produce a distinct unbalance of power, with an enormous advantage on the side of the democracies. I am not interested in balance of power, for it inevitably leads to war. I am interested in a distinct unbalance of power, so that the initiative is on our side.

If we had a powerful federal union side by side with another powerful federal union such as the U.S.S.R., then the weaker of the two could hardly attack, and the stronger of the two would not need to, and some years of peace might be secured during which a number of things might happen on the favorable side.

Some persons object to this proposal because they say it will lead to war by the most direct route. It might; the probability may be better than 50 per cent, but I maintain that any other alternative also has an enormous risk of leading to war.*

A "Federal Union of the Free" is advocated by Federal

---

* Harold C. Urey: "An Alternative Course for the Control of Atomic Energy," Bulletin of the Atomic Scientists, vol. 3, no. 6, pp. 139-142.

Professors Philip Morrison and Robert R. Wilson of the Physics De-

Union, Inc., of which Clarence K. Streit is president. A condensed statement of its proposal follows:

We believe in a Free World Republic as a final goal but deem it dangerously impractical to try to change all the United Nations together into a world government now. We find it no less impractical to trust to the UN or to the U.S. alone to secure liberty and peace. We support the UN as being about as strong an organization as can now be made of nations as divergent as the U.S. and the Soviet Union. But we believe it unwise to attempt now to abolish its veto, and folly to rely primarily on the UN for peace.

We favor founding now around the Atlantic a Federal Union of the Free composed of the few democracies that have proved their ability to guarantee individual liberty. We would create it as a nucleus designed to grow into a world republic by the peaceful extension of its free federal principles to others as rapidly as this proved practical.

We believe that by organic union such democracies as the United States, Britain, Canada, Australia, New Zealand, South Africa, Eire, France, Belgium, Holland, Switzerland, the Scandinavian countries, and the Philippines would gain such colossal moral and material power as to prevent dictatorship from attacking them or anyone.

---

partment of Cornell University oppose Dr. Urey's plan. They say it "simply can't be made to work."

"The United States of America has now the monopoly of the bomb, and a near-monopoly of heavy bombers," say these Cornell professors. "We are producing more of all the materials of industry than is the rest of the world together." But we do not now "feel security from attack. Nor would we behind an alliance, called a limited world government if you please, in which only those with whom we now have no differences would be joined. . . . Separation into half worlds would be the surest and most direct path to the all-out arms race, to the final stupid waste of another world war."[5]

John Fischer has expressed his belief that we can have two worlds [or two half-worlds] at peace with one another.

Thus we would solve the problem of securing peace long enough for the most divergent peoples to evolve to the point where universal union became practical. Meanwhile we would have this nuclear World United States put the authority of its immense and ever-growing power behind the UN, as a member of it.[6]

A World Constitutional Convention is now proposed by at least three groups—one in England and two in the United States. The group in England is composed of Mr. Henry Usborne, a member of the House of Commons, together with about 100 or more other members of Parliament who endorse his ideas. This group believes that it is not now politically possible to amend the Charter of the United Nations without destroying the organization, since the great powers cannot at present agree on the amendments necessary. "If world peace is to be maintained permanently," they say, "a World Parliament must soon be created which controls a monopoly of armed force. . . . This change cannot be achieved without war unless a Charter of World Government is drawn up showing exactly, and in detail, how the World Government will work."

There should be held in Geneva in the autumn of 1950, Mr. Usborne and his group believe, a "World Constituent Assembly" to be attended by representatives of all peoples of the world, one per million, from every country. This Assembly would draft a world constitution.[7]

In the United States, a Peoples World Constitutional Convention is proposed by World Republic, Inc. This organization believes it is necessary that the people have a federation of the countries of the world responsible to

Proposals for World Government 125

them and to them alone. A World Constitution would be submitted to the various governments for ratification by the people according to the laws of each country.[8]*

The actual drafting of a world constitution has now been underway since early in 1946. The "Committee to Frame a World Constitution" has undertaken the task under the leadership of Chancellor Robert M. Hutchins of the University of Chicago. While Mr. Usborne apparently believes that a world constitution can be drafted by an assembly of 2,300 or 2,400 founding fathers in a reasonable time, the Committee to Frame a World Constitution has proceeded on the belief that the task requires laborious effort

---

* "We cannot have world government . . . in any meaningful sense, without Russia," asserted Grenville Clark before the New York City Bar Association. Mr. Clark believes that a determined, persistent effort should be made to achieve world government that will provide an over-all settlement with Russia.

"What shall we ask of Russia? The indispensable things are, I submit: (1) her consent to explicit Charter amendments abolishing all military forces and armaments, except such as are actually required for internal order; and (2) her consent to futher amendments that will make the United Nations a genuine world federal government authorized to function effectively in the limited but vital field of war prevention."

After discussing the far-reaching implications of these provisions, Mr. Clark explains that "from the Russian standpoint, these would be immense concessions"—so great, that it is hopeless to expect her consent except through fair recognition of her stature in world affairs and of her legitimate vital interests.

"What are those interests and how can they be fairly met? The two greatest issues, he believes, are "the problem of the Dardanelles and Middle East oil." In respect to the Dardanelles, the solution, Mr. Clark says, "calls for placing all the great passages between seas and oceans under a reformed United Nations."

In respect to Middle East oil, he does not believe that Britain and the United States should expect to continue in virtually exclusive control, denying any share to the Soviet area. Here he explains, "the remedy lies in an imaginative, cooperative outlook whereby the Soviet sphere is allocated a fair share on fair terms."

On our part, he concludes, "we must shed much of our massive self-righteousness. On the Russian side, they must discard much of their obduracy and rigidity . . . There must be with us, as with them, at least some spirit of give and take."[9]

over a period of many months. Consequently, the Committee opened an office in February, 1946, and recruited a staff of approximately twelve employees, including scholars, administrators, and clerks.

"We of the Committee to Frame a World Constitution hope to be ready long before 1950," asserts Dr. Hutchins. "We do not think, of course, that our preliminary draft will be the law of the United World. We trust nevertheless that the tentative result of a collective effort of years will not be in vain. The world at large will have ample occasion to learn from our success and failures and to teach us and others. The Federal Convention in Geneva, or wherever else it may be, will not have to break through a wilderness of immature and contradictory proposals. A pattern will be available. We do not think it will be adopted; we dare to hope that it will not be ignored.

"Time is of the essence. It is now or never. Indeed straight is the gate and narrow is the way which leads to life. The obstacles in the way of World Government and the unity of mankind are staggering. But the other way lies perdition."[10]

We have set forth in this and the previous chapter various ways of building a united peaceful world. One of them is now in operation—the United Nations. The other plans presented are advocated by influential committees, nationwide in scope. We may now ask the question: What can individual citizens and local groups do to arouse interest in the present crisis and the winning of the peace? Several answers to the question are given in the following pages.

· 15 ·

Local Groups in Action

O N A SUNDAY AFTERNOON, SOME MONTHS AGO, A LARGE audience filled the central hall of the Enoch Pratt Free Library of Baltimore. The adjoining reading rooms were jammed and people hung precariously over the balcony railings above. The library staff planned for an attendance of 1,200 to 1,500 persons; 3,422 came. The occasion—a meeting on the problem of atomic energy. On four following Sundays a second auditorium was wired for sound to take care of the overflow.

Conferences between Emerson Greenaway, Librarian, and staff members resulted in a conviction that this problem "must be brought home" to the local community.

These librarians decided not only that the best books on the subject should be made available, but that they would in addition secure and utilize speakers, motion picture films, and an "exhibit."

The aid of the Maryland Academy of Sciences and of Johns Hopkins University scientists was secured and a program of wide publicity was developed. Hundreds of local organizations cooperated and distributed over 50,000 book lists and a like number of leaflets.

In store windows 3,500 posters and placards were displayed. Newspapers, school bulletins, and house organs cooperated and interested citizens subscribed directly or indirectly the sum of $3,000.

The speakers, who appeared at the five Sunday afternoon meetings, included Leland Stowe, foreign correspondent; William Higinbotham, Executive Secretary of the Federation of American Scientists; Vera M. Dean of the Foreign Policy Association; Norman Cousins, editor of the *Saturday Review of Literature*, as well as prominent local speakers.

Four motion picture films were selected. One was "A Tale of Two Cities," produced by the Signal Corps of the United States Army. The other three were cartoon strips prepared by the National Committee on Atomic Information—"How to Live with the Atom," "World Control of Atomic Energy," and "Up and Atom." These films were shown on four successive Tuesdays at 12:30 for the benefit of downtown shoppers and office workers.

The exhibit consisted of 21 "free-standing lighted panels" bearing charts, cutouts, photographs, photostats, and other visual aids. The panels were 7 feet high, ranging in width from 2½ to 5 feet. They portrayed the confusion of man regarding atomic energy and its implications, how atomic energy was discovered and developed, and what it can do as a force for destruction and how it can serve human progress.

The National Committee for Atomic Information* has taken over this exhibit and is routing it throughout the country. It may be secured free of charge, except for transportation costs from the preceding place of exhibition.

* The address of this Committee is 1749 L St., Washington, D. C.

Local Groups in Action

The American Library Association feels strongly that libraries throughout the country would do well to follow the example of the Enoch Pratt Free Library and its educational work regarding atomic energy.

This library has proved that "enthusiasm has its own chain reaction." Individuals may not have the resources to put on as large-scale a program as did the Pratt Library. But, as Mr. Emerson Greenaway, the librarian, has said, "The main thing is to go ahead and do something!"[1]

When the United World Federalists was organized as a result of a merger of six national agencies, one of the groups brought into the new and larger organization was Student Federalists. Here is the story of its origin told by Harris Wofford, Jr., who was responsible for the first local group in Scarsdale, New York.

One night in 1941 our family settled down to a normal weekday evening [writes Wofford]. Upstairs I was trying to concentrate on stiff tenth-grade homework assignments in Latin and algebra. In the usual custom of my generation, I had the radio on full blast. I was 15.

As my Latin grew intolerably difficult and Mr. District Attorney became increasingly interesting, I put homework aside and took a bath, turning the radio up a little louder so I wouldn't miss anything.

Mr. District Attorney rushed to its dramatic finish. I was still in the bathtub when the program changed and speeches started coming over the air from a banquet at the Waldorf-Astoria. If the radio had been within reach I would have quickly turned the dials. But I had to listen.

The subject was "A World Federal Union of Free Men."

The speakers—Frank Kingdon, Clare Boothe Luce, Clarence Streit, Thomas Mann, and Dorothy Thompson—all advocated Union Now. I found myself listening attentively to proposals of a Union which "would expand around the world until it became the United States of all mankind."

As I heard these words, the idea of the United States of the World *in our time* changed my life's course.

In the morning I wrote a letter to the sponsors of the broadcast, Federal Union, Inc., formed by proponents of the principles set forth in Clarence Streit's *Union Now*. I asked for two things: information, and what could I *do*? Point 1 was answered by printed material which I read carefully and approvingly. Point 2, unanswered, was apparently up to me.

I had been discussing World Union with a number of friends in school. It struck home with the same initial clarity to many of them. Several of us went to New York, marched into the Federal Union, Inc., office and asked, "What can students do to help?" The receptionist looked doubtful. Finally she said, "Well, why don't you form our only high school chapter?"

It required seven members to be chartered as a chapter. That night Mary Ellen Purdy and I got on our bikes, armed with an idea and a bundle of literature, and rode out to enlist the seven members. We both missed supper but together had signed up eight charter members. That night the Scarsdale Student Chapter of Federal Union was born. Several weeks later, on March 23, 1942, our official charter came through.

At the end of the year, many members of our chapter graduated, but in the fall we enlisted others. Howard Mor-

Local Groups in Action 131

gan, probably the most popular leader in the class, joined and became more and more active. We spoke before every history class in school that fall. We held our first assembly, and frequent afternoon open forums.

Then we had a speaking team. Coburn Ayer, our English teacher, had coached a few of us after school on public speaking and we were soon traveling all over Westchester County on our bicycles, speaking from town to town to high school assemblies and forums.

Perhaps this sounds easier than it was. Looking back over the progress we made that winter, I remember how hard the progress came, how much opposition we faced. One classmate publicly campaigned against us. "Union Never!" he would shout when he passed a Student Federalist in the halls.

In early 1943 came the great news that out in Minnesota a high school debate champion named Thomas Hughes, entirely on his own, without knowing that in the East we had done the same thing, formed a High School Chapter of Federal Union! Soon after that Jerry Miller in Sarasota, Florida, wrote us that he too was forming a chapter, spearheading our work in the South.

A year from the time we had first started in Scarsdale, there were about 150 student members in a half dozen states and a "national" office in Scarsdale which was mimeographing a new paper called the *Student Federalist*.

In April, I resigned as Scarsdale chapter president to concentrate on national work as chairman of the Student Federalist Council.

By the end of the summer we had raised just under $500 to inauguarte a *nationwide movement*. Together we Student Federalists started tackling the job, thinking we were on the home stretch towards World Union. Actually we were only rounding the first easy lap of a long, difficult journey.*

Oak Ridge, Tennessee, high school students are close to the problem of atomic energy, living as they do, almost in the shadow of huge factories making atomic explosives. After an assembly talk by an Oak Ridge scientist, the Youth Council on the Atomic Crisis was organized, and soon, in 1946, there were 200 members. They pledged themselves to learn the facts on the crisis and promote discussion among family groups, relatives, and associates. They published an extra-large ten-page issue of their high school paper on the subject, and distributed over 12,000 copies. Within a year the Oak Ridge organization was affiliated with YCAC's in twenty-three states. More recently, Oak Ridge representatives have participated in many educational meetings of various kinds. For instance, they have had speaking engagements in at least fifteen states and have appeared on the programs of five or more national conventions.[2]

In Charlottesville, Virginia (the site of the University of Virginia), an "Atomic Energy Week" was observed in 1947. On Sunday, May 11, sermons on some aspect of atomic energy were preached in the various local churches, on Monday there was a meeting in the Lane High School, at which two nationally known radio commentators were the speakers. From Sunday through Wednesday a full-length Hollywoood movie, "The Beginning or the End," was shown in one of the theatres.

* Condensed from *It's Up to Us*, by Harris Wofford, Jr., Harcourt, Brace and Company, New York, 1946, with the permission of the author and publisher.

Local Groups in Action 133

Each day there were free showings of sound films in a vacant store to reach the "man on the street." Throughout the week there were radio programs and announcements, exhibits in store windows, and the constant circulation of films, with speakers, among schools, civic groups, and church groups.

On Thursday, an institute was held at the University, in which outstanding political scientists and nationally known atomic scientists participated. There were an afternoon and an evening session.

The program was made possible by contributions from banks, educational organizations, theatres, men's clubs, women's organizations, and business establishments.[3]

The Charlottesville enterprise was only one of several in Virginia. In eleven other cities of the state one day meetings were held. Furthermore, Virginia is only one of five states in which conferences have been arranged by the Association of Scientists for Atomic Education. They have been held, in addition, in the principle cities of South Carolina, Florida, Georgia, and Louisiana.

In general, two atomic scientists have appeared on the program with two social or political scientists from the university of the state. The need for continuing education, and the readiness of local citizens to do the necessary work, was evident in lively discussion periods which followed the lectures.[4]

These are stories of individuals and groups in the United States that have decided "to go ahead and do something" through organized effort, to arouse public opinion regarding the threat of atomic warfare and the necessity for a world organization strong enough to prevent war. Other individuals and groups, here and abroad, stirred by reports on the plight of peoples in devastated areas, have done something to alleviate their suffering, usually without the backing of organizations. A few of these expressions of international friendship are now briefly described.

· 16 ·

Beginnings of World-wide Friendliness

D-D DAY, ON THANKSGIVING DAY 1946, MARKED THE culmination of the efforts of Dunkirk, New York, to help the people of Dunkerque, France, who had taken a severe beating during the occupation of the Germans. [Quentin Reynolds, after spending a week in Dunkirk, tells the story.] The firehouse of the New York town had been selected as a warehouse for goods to be donated to the French community. Clothing, canned goods, medical and surgical supplies kept piling up, as well as dental equipment, all sorts of farming implements, and toys, blankets, cooking utensils. Soon the big red engine had to be moved out. Then a herd of breeding cattle, ten young heifers, two bulls, a dozen goats and a dozen pigs, which could not conveniently be kept in the fire house, were taken to a nearby farm.

Eighteen members of the Hi Ya Club saved enough money to buy and contribute a microscope; a local union at one manufacturing plant got together to contribute fifty dozen shovels. School youngsters brought in hundreds of pencils and school supplies.

When D-D Day arrived, French Ambassador Henri

Bonnet came to accept the gifts. There was a parade with the Ambassador, Mayor Walter Murray, and Charles Boyer (the motion picture actor) on the reviewing stand. Eighteen big trucks went by laden with gifts for Dunkerque, Dunkirk's sister city. The local committee had undertaken to raise $2,500; the value of the goods collected was about $100,000.

Many things happened that day that the people still talk about. But the really big thing that happened is a thing that none of them can explain. "It was not just a momentary flash of generosity quickly forgotten," writes Quentin Reynolds. "Something happened that seems to have left a permanent mark on every man, woman and child in the city. Dunkirk grew up; Dunkirk realized it was not just a pleasant little city on Lake Erie. Dunkirk realized that it was a part of the world."

The excitement of D-D Day had hardly died down when John McCauliff, a businessman, saw an item in a Buffalo paper regarding the desperate condition of the Polish people. "If we helped France, why not help Poland?" he asked. Many persons in Dunkirk had relatives in Poland. A drive for funds was launched, and goods and money to the extent of $150,000 were soon collected and sent to Poland.

People began to notice a new spirit in their relations with their fellows. Absences from the high school were virtually eliminated. Students became more eager to gain knowledge of current events. Labor troubles seemed to disappear. Better interracial understanding developed. Three competing veterans' organizations got together and formed a central council. The Chief of Police said he was virtually out of a job. "If you ask me what happened here," said John

Beginnings of World-wide Friendliness

McCauliff, "all I can say is God smiled on us and made Dunkirk the best damned city in America."

"Dunkirk was tired of reading about diplomatic squabbles, and about United Nations debates," continues Mr. Reynolds. "Dunkirk, without being able to put it into words, was embarking on a program of direct contact with the people of the Old World, with no middleman in the form of government or diplomat involved." Hundreds of letters of thanks came from Dunkerque, and when they were translated and printed in the local paper the people of Dunkirk said thoughtfully, "Why, these are our kind of folks. They're no different from us. Maybe people all over the world are just like us. If they are, they don't want any more wars."

Wally Brennan, editor of the *Dunkirk Observer*, pushed his typewriter aside and leaned back in his old chair. "The governments, our government and the governments of all the other countries," he said, "are doing a lot of complicated things in the way of debates and treaties and agreements. But it doesn't seem to be getting the world any nearer to a permanent peace. Just suppose the people of all the different countries somehow got together and got talking to one another without any diplomats interfering? The governments of the world might find out how unpopular war is."

As a result of considerable discussion, a Dunkirk Society has been organized. "Suppose a few hundred other cities in America who feel as we do were to do what we did?" asks editor Brennan, "sort of adopt some place in Europe and help it along. You know, it might make the people of Europe think pretty highly of us as a country. It might make them realize that even though we talk different languages, we're the same kind of folks."

Again, at Thanksgiving time in 1947, Dunkirk (assisted on this occasion by Fredonia) contributed $100,000 in supplies and cash, in a "hands-across-the-sea" gesture for the rehabilitation of Anzio, Italy where many American youths fell in the last war.*

Many communities, local groups, organizations and individuals, in various ways, have given generously of time, money and goods to assist the peoples of war-stricken areas in Europe and Asia.

The Save the Children Federation obtained sponsors in the United States for more than 1,000 schools in the devastated countries of Europe. These sponsors see that the school children's "three greatest needs are met: supplementary food, clothing and shoes, and books, pencils and paper."[1]

Harvard had sent $40,000 worth of food to Europe's starving students, and then an Austrian-born Harvard man, Clemens Heller, decided it was time to do more. Money was raised and a dozen educators journeyed to Salzburg, Austria, in 1947, to establish a summer seminar; 100 persons, mostly teachers, attended. The American professors received no salary; some of them even paid their own traveling expenses.[2]

High school students at Manchester, New Hampshire, saw a newspaper item about a war-ravaged European school and decided to raise funds to buy new equipment. When

* Condensed from "Dunkirk—One World Town," by Quentin Reynolds, Colliers, June 7, 1947. With the permission of the author and publisher. Paragraph on the Anzio project is from the New York Times, November 28, 1947.

Beginnings of World-wide Friendliness

the drive ended, $800 was available. At Elizabeth Irving High School and the Little Red School House in Greenwich Village, New York City, students took on the responsibility of outfitting boys and girls in a public school in Florence, Italy. The American Youth for World Youth, a federation of 3,000 school and camp organizations, makes recommendations as to what articles may be sent to young people abroad. Each gift is accompanied by a personal letter.[3] The American Red Cross and other agencies promote correspondence with young people of other lands.*

Class Memorial Funds are being used to provide books and other needed materials for students in war-torn lands. For the school year which ended in June 1947, 140 graduating classes had built up such funds.[4]

Deep in Finland's pine forests, a little community of 350 farmers and their children welcomed thirty-five visitors sent in 1947 by the American Friends Service Committee. These visitors, from 18 to 28 years of age, lived in tents and worked each day clearing the land, draining the swamps, and fitting together the sturdy log houses that would soon make the community a real town.[5] Twenty-five French boys, aged 17 to 22 years, arrived in the United States in July, 1947, to participate in Quaker work camps and international seminars. Later in the summer, they were entertained in American homes, and in October returned to France.[6]

The Society of Friends of the United States and Britain was awarded the Nobel Peace Prize for 1947 for friendly services in the cause of peace (only a few of which are recorded in this small volume). The American Friends Service Committee alone has administered $60,000,000 in

---

* A list of these agencies will be found on p. 183.

more than twenty devastated countries and is still carrying on its efficiently organized work. "Such work," states the New York *Times*, editorially, "is the very foundation of peace and understanding."[7]

The International Committee of the Y.M.C.A.—after maintaining, during the war, 600 camps for prisoners, civilian internees, and others in twenty-eight countries—is now raising $8,000,000 to rebuild or repair 105 Y.M.C.A. buildings in war-torn areas. After his trip around the world Wendell Willkie reported that missionaries were the most popular foreigners in every country he visited.[8]

The Experiment in International Living, Inc. for several years has been bringing groups of Mexicans and Guatemalans to the United States to live in families; and in the summer of 1947, 145 American students went abroad to live with families in Norway, Sweden, Denmark, Mexico, and Guatemala and to work in camps for underprivileged children in France. The largest number, about ninety, went to France.[9]

The International Ladies Garment Workers Union presented to the Netherlands' ambassador in March, 1947, a check for $100,000 as a loan to the Netherlands Trolley and Railway Workers Union for the purchase of overalls, shoes, and other work clothes.[10]

The Medical and Surgical Relief Committee has sent more than $1,000,000 worth of medical and surgical supplies to needy areas overseas. In October, 1947, the Committee was planning to continue its work.[11]

Finally, we should not forget the "friendship train," promoted by Drew Pearson, columnist, which started from California in November, 1947, and reached New York in several sections with a total of over 500 cars of grain and other foodstuffs for the French and Italian people—a symbol of the real concern of the American people for their less fortunate neighbors in these countries. This fine enterprise stimulated similar projects. The "Abraham Lincoln friendship train," assembled by the Christian Rural Overseas Program, was made up of over 240 carloads.[12]

European peoples have made fine sacrifices to assist their less fortunate neighbors. The Henri Dunant Center of "Relief for Children," a Swiss Red Cross project, originated, staffed, and financed entirely by Swiss volunteers, has cared for more than 100,000 children since 1942 from twelve European countries. Undernourished boys and girls between the ages of 4 and 14 have been brought to Switzerland for a three months' period of recuperation. From the reception centers they are sent to live with Swiss families. The before-and-after appearance of the children—from rags and hunger-pinched faces to warm clothes and well-fed bodies—is a sight to gladden any heart.[13] The Swiss people in the spring of 1947 were also building a Children's Village at Trogen. Nearly half of the contemplated twenty-four houses had been erected, each accomodating approximately sixteen children. Boys and girls from France, Poland, and Holland were living in the new community, which it is hoped will become a self-supporting village.[14]

Norwegians, during the summer of 1947, brought hundreds of European children to their country to enjoy the serenity and beauty of the fjords, mountains, and woods. They were housed and well-clothed and fed back to health. From Czechoslovakia came 300, from Poland 260, from Austria 60. All were brought to Norway through the efforts of the Norwegian Political Prisoners Association, a group whose members had been interned in Nazi camps. Norway was one of the first nations to be occupied by Germany and the country is not wealthy. But the Norwegians, some time ago, asked CARE to "divert food and textile packages to other areas of Europe where privation is acute."[15]

Here, in limited space, we have been able to consider only a few of many modest, but generous, efforts by organizations and individuals, at home and abroad, to develop world-wide friendliness.

The good will of the people of the United States has expressed itself in government action through the Marshall Plan. To this major effort we now give our attention.

· 17 ·

The Marshall Plan
—An Adventure in Friendliness

Secretary of state george c. marshall on june 5, 1947, received an honorary degree from Harvard College. In his acceptance speech, he said (1) that Europe will be short of food and other necessities for three or four years; (2) that, before the United States provides such necessities, European countries should find out what they can do for themselves and one another, (3) that Europe should plan for itself, and (4) that, when Europe has done that the United States should give what friendly aid it can. Upon inquiry, Secretary Marshall indicated that he had in mind all of Europe including both Soviet Russia and Great Britain.*[1]

Eleven days after Mr. Marshall's speech, Ernest Bevin, the British Foreign Secretary, flew to Paris to discuss the speech with Georges Bidault, Foreign Minister of France. These two decided to seek Russia's participation in the

---

* It should be noted that Mr. Marshall did not propose a plan. He merely suggested that European countries develop a plan for the consideration of the United States. Thus, there really is no "Marshall Plan." Because of accepted usage, however, we shall continue to use the term.

preparation of a note to Secretary Marshall. Mr. Molotov in behalf of Russia agreed to a conference which was held on June 27th. Soon, however, Mr. Molotov found himself unable to agree with Secretary Marshall's proposal.

Subsequently, sixteen nations sent representatives to a conference in Paris who agreed upon a report, now referred to as the E.R.P. (European Recovery Program). The program provides that each country: (1) modernize its equipment and increase its production especially in agriculture, fuel, power and transportation; (2) create and maintain internal financial stability; (3) develop economic cooperation between the other European counties; (4) try to find a way to pay their debts to the United States by exports from Europe or elsewhere.[2]

During the autumn, competent committees made careful studies of the plan and submitted reports to the President. Promptly, the Congress appropriated $540,000,000 for an "interim aid program" for Austria, France and Italy.[3] Then on December 19, 1947 the President proposed to the Congress the appropriation of $6,800,000,000 for the period ending June 30, 1949 and a total of $17 billions for the four and one-quarter year period ending June 30, 1952.[4] The graph on the opposite page shows the average amount per year requested for this period, compared with the cost of defense for one year and the estimated personal income of the nation for 1947.

Immediately the debate got under way, first among newspapers and periodicals. The Christmas holidays were available to members of Congress for study of the plan. Upon their return to Washington early in January hearings began in House and Senate committees.

Week after week the debate continued—a debate without

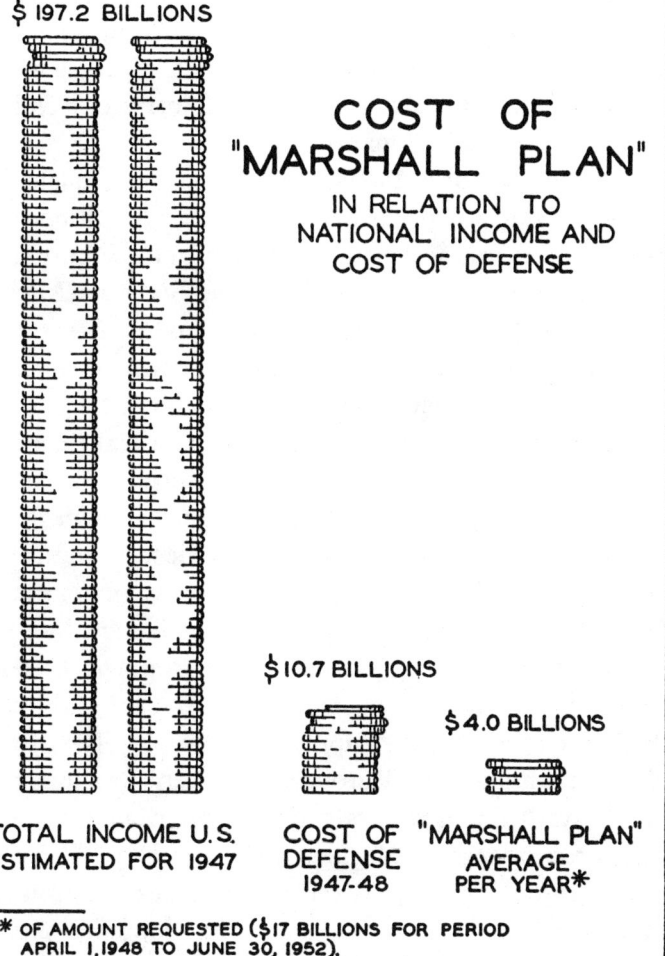

Figure 3

parallel in history. Only Eskimos, the dwarf blacks of the Congo and a few other primitive peoples, wrote *Collier's*,[5] were not concerned. It might be regarded as a part of the great debate now going on all over the world—a debate, suggested Frederick H. Osborn, between ideals, a debate whose outcome will decide the kind of world our children will live in for generations to come.[6]

From the beginning the issue was clear—clear at least to those who understood what was happening in Europe. The United States must provide adequate aid to Europe, or misery and disorder would increase. Said Secretary Marshall at a joint meeting of Senate and House committees, "The Congress will be called upon to make decisions which, although less spectacular and dramatic, will be no less important for the future of our country and the world than those of the war years. . . . The automatic success of the program cannot be guaranteed. . . . The risks are real. . . . They are, however, risks which have been carefully calculated, and I believe the chances of success are good. . . . We must not fail to meet this inspiring challenge. We must not permit the free community of Europe to be extinguished."[7]

In order to keep the "supply pipeline" to Italy, France and Austria filled until European Recovery Program funds became available, Congress provided on March 29, 1948, $55,000,000 of additional interim aid.[8]

Then on April 2, 1948, after extensive discussion in both House and Senate, Congress acted. It adopted a measure providing a total of $6,098,000,000, including the following: $5,300,000,000 for use by the sixteen European nations during the period of April 1, 1948 to April 1, 1949; $338,-000,000 for economic aid to China; $125,000,000 for China

The Marshall Plan—An Adventure in Friendliness

which can be used for military purposes; $275,000,000 for Greek-Turkish military aid; and $60,000,000 for the International Children's Fund of the United Nations. Action on the balance of the 4¼ year period covered by the original European Recovery Program was postponed, but a moral commitment for this period was retained.[9] Thus the United States decided to provide substantial assistance in the economic rehabilitation of Western Europe, and in relieving three emergencies.

Here is a major enterprise in friendliness. "The most unsordid act in history," the London Economist called the Marshall Plan. While it may reveal enlightened self-interest, wrote the editor, "the degree of enlightenment is so infinitely in advance of any yet shown by a great nation in time of peace that it may well be questioned whether it comes into the category at all." The executive committee of the Federal Council of Churches described it as "one of history's most momentous affirmations of faith."[10]

The measure thus adopted will probably be effective. That is, it will tide over European economy for about a year. It should do more, if trade relations can be reestablished between Eastern and Western Europe. If the original four and one-quarter year plan is implemented by adequate additional appropriations, it should make Europe stronger in resisting Communism—at least until Russia can make atomic bombs. It will not stop the present armaments race which consumes over $10 billions per year by the United States, and a large but probably lower amount by Russia. (The annual military budgets of all nations, we may remember, now add up to $30 billions[11]). It will probably not prevent the indefinite continuation of the "cold war," between these two countries. The Marshall Plan in itself will not bring lasting peace. But if the plan is administered successfully, and promoted so that our people will make the sustained effort the original program calls for, then it can be honestly justified as the incentive to the unification of Europe and as an investment in peace.[12]

Are the limitations of the Marshall Plan due to caution and a lack of vision? We have been generous, especially toward Western Europe. Isolationists and others say we have gone too far. Is it possible, however, that we have not gone far enough? While this nation is the wealthiest in the entire world, we appropriated for the European Recovery Program for the 12 months' period ending April 1, 1949 less than three per cent of the nation's estimated personal income for 1947;[13] and, as shown by the graph, we have been spending annually on military force over 2½ times as much as the amount per year originally requested for the development of the program.

Might we make greater progress toward a peaceful, united world in developing a long term policy, were we to place less reliance on force and more on generous friendliness? This question we will now consider.

· 18 ·

Force and Friendliness

IN TWO RECENT CRISES THE UNITED STATES HAS DEPENDED on military force to preserve democracy. With our allies we have won two world wars, and in large important areas, we have successfully defended democracy. In an increasing number of regions, however, democracy is now being destroyed. We have not succeeded in "winning the peace." Can it be that force has failed?

Failed, even though at present—and perhaps for a long time—force may seem necessary?

Can we depend on force to establish permanent peace?

The expenditure on armaments of the vast wealth of nations is bearing us closer and closer to another cataclysm. At times and in particular places the world situation seems encouraging, as in Italy and India early in 1948. But in respect to Greece, China, Korea, Palestine and Germany, and in regard to the threat of conflict in Scandinavia, Iran and Turkey, are not prospects for the future darker than they were at the end of 1946?[1] Still more discouraging is the growing lack of cooperation in the United Nations.

There is not only a general trend toward catastrophe, there are also specific dangers of accidents and "incidents."

Already there have been minor incidents on the Russian-United States border in Germany. The consensus in informed quarters in Washington, writes Ernest K. Lindley, is this: "No war, unless the Kremlin miscalculates or one of its puppets goes wild."[2] But who knows that the Kremlin will not "miscalculate," or that a puppet state will not go "wild"? John Foster Dulles, adviser to Secretary Marshall, while seeing no immediate danger of a shooting war, realizes that both Russia and the United States assume "that the other fellow has great self-control."[3] Is the assumption warranted?

There is a sentiment among some leaders in favor of a "preventive war." If we must have war with Russia, why not have it now while we have a monopoly on atomic bombs? If such an attitude is entertained by responsible military officers, may all the precautions desirable be taken to prevent accidents? A conscious or a subconscious attitude may exert a controlling influence in a critical situation.*

There is real danger; but war talk, all too common these days, may accentuate the danger. Neither the Russian nor American people want war, writes David Lawrence, well known Washington editor; but to refuse to prepare for war is to court disaster. On the other hand, to argue that military preparedness prevents wars is an illusion. Standing armies in Europe have helped postpone wars, he explains, "only until another nation could build up a rival army or its equivalent in naval power." While military preparedness is absolutely imperative "when there is no other kind of preparedness to prevent war," is America, he inquires, "bankrupt of intellectual ideas that could prevent war by any other means?"[4]

* See Chapter VI, p. 46 ff.

Force and Friendliness

The philosophy of might is failing. The shrewd bargaining of the old diplomacy no longer works. Because the atom bomb has altered profoundly the nature of the world, observes Dr. Einstein, the human race "finds itself in a new habitat . . . Our defense is not in armaments, nor in science, nor in going underground. Our defense is in law and order." He emphasizes that "a new type of thinking is essential if mankind is to survive and move to higher levels."[5] A businessman agrees. "Whenever a civilization is faced with a new and fundamental challenge," asserts W. T. Holliday, President of the Standard Oil Company of Ohio . . ." its men and women must change their habits of thought . . . or they go under."[6] As Norman Cousins so well explains, "modern man is obsolete."[7] This is especially true in international affairs.

While, for the present, we must continue to rely on military might, has not the time come for us at least to consider a different way to settle disputes between nations? To win the peace may the bold use of friendliness and faith be more successful, friendliness and faith and the spirit of adventure without which, says philosopher Alfred North Whitehead, "civilization is in full decay"? Advance, he asserts, or we go backwards. "The pure conservative is fighting against the essence of the universe."[8]

Can we help Russia now? She apparently does not want help. At present, force seems necessary; prompt and decisive force may be required.

In the meantime, is it too late to try friendliness, faith and the spirit of adventure in other parts of Europe and Asia? It may be too late. Citizens of France saved today, may be bombed into oblivion tomorrow. Soon the situation may become so much more critical that the only course remaining is to fight it out—fight it out with the most destructive weapons the mind of man has ever invented. It may seem too late. But, with the hope that there is time, and with a growing conviction that force is not the ultimate solution—that it will never bring lasting peace—let us consider briefly what other ways may be tried.

The setting up of a world government, one of the specific ways proposed, would in itself be an act of faith. Atomic scientists and others who recommend the creation of a world government have no assurance it would prevent war. Dr. Harold C. Urey said in January, 1948, he was not sure world government would not lead to war with Russia. "Maybe it will," he declared, "but the alternative—to do nothing—certainly will."[9]

Mankind is in the predicament of an outcast who is lost, and stands on the edge of a canyon, suggests Mr. Cousins. "Behind him rages a forest fire, drawing ever closer. In front of him is a sheer drop of several hundred feet. But the gap across this canyon to the other side is only ten feet wide. Ten feet! He has never jumped ten feet before. He has no way of knowing that he can jump it now. He has no choice but to try . . . Mankind today is involved in a somewhat similar predicament . . . there is no time to build a footbridge. It happens that we cannot take just a step forward, but must jump . . . Although world government provides a better method and a better chance of preserving world peace than man has ever possessed, it cannot provide a guarantee of world peace."[10] Conflict draws ever closer. Ultimately, we may have to take the chance, exercise faith and boldness, and leap forward.

Force and Friendliness

Our nation, in its participation in the work of the United Nations, is in a position to utilize greater friendliness and faith. We were not willing to implement the Marshall Plan through the United Nations, but we will have other occasions to support the United Nations.

A vast amount of relief and rehabilitation is needed in Europe. The American Friends Service Committee has trained young people in the methods, and has sent them to Europe to go into ruined villages and out on her wasted farms to work alongside European youth in clearing away ruins and getting the people back on their feet. These efforts have met with fine success. Why should not our government profit by this experience? Such work, asserted the *New York Times*, "is the very foundation of peace and understanding."* Here is one enterprise that could well be implemented through the United Nations.

We might, perhaps, supply the capital, the engineers and equipment, for a Jordan River Authority in Palestine, suggested John Fischer,† and a Tigris and Euphrates River Authority in the Arab lands to the east. Herbert Hoover, it will be remembered, has already described how the job could be done. American machinery and know-how could save 20 crucial years in the long-overdue industrialization of India and China. And if humanitarianism were combined with industrialization, consider the number of lives that might be saved! Here is a proposal that requires faith—and also money.

Did we not agree with John Fischer when he asserted that "our economic strength is the greatest single advantage we have in the contest against the Communist half of the

* See p. 140.
† See p. 76.

world?" He does not believe "that we are fools enough not to use it where it will do the most good." Here are major enterprises that might be implemented through the United Nations, with the aid of the International Bank for Reconstruction and Development ("World Bank"), in such a way that they would be more effective as cooperative undertakings than if developed through independent action by the United States.

There are practical reasons why the United States should initiate various venturesome projects of helpfulness. Let us be frankly realistic. First, in the same way as we needed the support of friendly nations when we entered and fought World War II, so we need friends now, in Western Europe, in the Near East, in the Far East and in South America—from both a diplomatic and military point of view. Second, we need these various nations as friends, to keep Russia from absorbing them into her sphere of influence. In some cases, this is an urgent need. We must think more quickly and act more quickly than Russia. We must "out-smart" Russia in winning friends. In doing so we must not be arrogant. Our great debt to Europe for her courage and sacrifice in the last war should be remembered. Yet we must be bolder than Russia. "We must show that the free societies can generate forces for construction," writes John Foster Dulles, which will render impotent Soviet methods of destruction.[11]

Eventually, Russia must be the major objective of our program of friendliness—a program that may call for aggressiveness. The rare patience of Secretary Marshall and his predecessors may have prevented violent conflict. But this patience, in a sense, has been a negative kind of thing. We have been neither bold nor adroit in our dealings with the Soviet government. "U.S. leadership in world affairs," said *Time* in March, 1948, except for a few notable exceptions, has been "unimaginative and uncertain."[12] While we have been generous, we have not acted with the audacity required. Our timidity has permitted the minds of the Russians, and of our own people, to become concentrated on war-like measures. Thus, fear looms large in our national lives as the motivating influence during these critical years.

If we were to act generously enough in regard to programs of relief and rehabilitation, and in regard to such projects as a Jordan River Authority, we might do more than improve conditions in backward areas; we might win from Russia a greater and more friendly respect than the display of military power will ever elicit. Turn her attention, it is proposed, from our threatening atomic energy plants to great rehabilitation and engineering projects! Such a change in Russia's outlook, if it can be successfully encouraged, might lead her United Nations representatives, and our own also, to be more willing to make the concessions necessary to agreement on atomic energy control. Then the proposal of the process development engineers might at least be considered—the postponement for 20 years of the large-scale production of nuclear fuel for use in making bombs and for *possible* use in producing power.* International control of atomic energy might then be more readily achieved with the cooperation of Rusisa; there would be no danger of seizure of our atomic energy plants for military use; we would then have far greater security (say these engineers); and a vast amount of money would be saved.

We do not know what influences may bring favorable changes in international relations. When Russia brought Czechoslovakia under her control, Joseph and Stewart

* See p. 84.

Alsop ventured the suggestion that the strain on Soviet administrative machinery, thereby increased, might later weaken Russia's grip and provide an opportunity for successful negotiation with the western nations.[13]

The possibility of reaching the Russian people directly should not be rejected. General A. G. L. McNaughton, Canadian delegate to the United Nations Atomic Energy Commission, thinks we will be able by some course to reach the people of Russia over the opposition of those who now control Russian policy.[14] The people are friendly, travelers have found; they want peace. And there are evidences of discontent within the Soviet Union. The United States Information Service, whose expansion was authorized in January, 1948, should be able, through the "Voice of America," the press and other measures, to promote better understanding. Some unexpected development of events may point the way to the attainment of this objective.

Let us assign the great task of winning the peace to an outstanding board of strategy of men with the imagination, faith, boldness, comprehension and global-mindedness of General Marshall and the others whose masterful strategy enabled us to win (in a military sense) the second World War—men competent to develop an equally masterful strategy in attacking the baffling problems of these momentous post-war days. And let us not be afraid of spending the money necessary, $15 or $20 billions, perhaps—not much considering the larger amounts past wars and preparedness are now costing us. Even if European peoples cannot immediately see that rehabilitation and construction projects are under way, the knowledge that plans are being rapidly and vigorously developed would improve their morale. If, fur-

Force and Friendliness

ther, they can be told that the United States is now assuming dynamic leadership in other respects, that fact would strengthen their resistance against Russian aggression.

Still more important, the desired effect on the Soviet government may become evident. Then, even while we are prepared to use force if need arises, let us make it clear to Russia that we are willing to help her, as well as the rest of Europe; that we are ready for friendly negotiations, in the various "organs," councils and agencies of the United Nations, or anywhere else. Until the United States and Russia do become friends, we can never have a united world.*

These various ideas may be "visionary." But so is the Sermon on the Mount; as an ideal we do not reject it. Must a great concept always be accepted only as an ideal, never as a basis for action?

Force has failed. While, paradoxically, we must now use force in dealing with emergencies, has the time not come when the other way must be carefully considered in the great hope that it may be tried, years from now perhaps, possibly soon—tried on a large scale as a great adventure in friendliness?

Civilization—let us not fool ourselves—has come face to face with a great dilemma, perhaps the most portentous in all history. On the one hand are nationalism, suspicion, fear, dependence on force and a course of action that is leading us on to suicide. On the other hand, there is the way —the necessarily hard and uncertain way—of adventure, faith and friendliness.

Cold logic, understanding and realism compel us to reject

* See again statement of Thomas K. Finletter, footnote, page 26.

the first as a means to enduring peace. Have we enough imagination to see merit in the second?

Let us postpone the answer until we have reviewed the relevant facts and other considerations briefly set forth in the preceding chapters. This review we now undertake.

· 19 ·

Mighty Enterprise

WE HAVE HAD A VIVID REPORT ON THE GREAT BALL of fire, and the mighty thunder and the trembling of the earth that accompanied the release of atomic energy on the New Mexico desert; we have read terrifying descriptions of the destruction wrought by two relatively small atom bombs, and of the devastation which even deadlier weapons may cause if war comes in the new atomic era.

We have hurriedly reviewed some of the facts of the postwar chaos, and we have observed that, while the economic and political situation in Europe and Asia is alarmingly acute, there is fear in Europe that the United States will suffer a grave depression; and Russia is betting that we will have one. We have seen that such a failure of American democracy would almost certainly lead to the rapid spread of communism throughout Europe—perhaps to the rest of the world. Russia, we have discovered, is, for us, the master problem. But the United States, in becoming the greatest military power in peace time history, is a problem to Russia. Now these two nations are engaged in an armaments race costing the peoples of both countries many billions of dollars per year—money urgently needed for food, homes, schools, and hospitals. Each nation is rapidly getting ready for World War III, a war that may bring about the downfall of western civilization. Already we have two worlds instead of the "one world" which has been our great hope.

A basic difficulty here at home, it has been explained, is our preoccupation with the physical sciences to the neglect of economics, sociology, government, and the other social sciences. Many do not understand international problems. We do not know why men fight. We do not understand Europe and Asia, and the poverty, fears, and frustrations of their peoples. We have exercised our inventive genius in the development of great mechanical power; we do not know how best to use it for the good of society. The major, long-term question before us is this: Are atomic energy, and machines and science in general, "to be used to degrade man and destroy him or to augment the dignity and nobility of human kind"? The adequate education of youth in the humanities and the social sciences may enable society eventually to meet this issue wisely and effectively —if we survive the immediate crisis.

On June 26, 1945, the United States signed the United Nations Charter with high hope that here was the way to peace. Now after 3 years we have lost much of that hope. If the UN is to succeed, the Charter, in the judgment of many individuals and various groups, must be revised and the UN given such power, legislative and military, as is necessary to assure peace. Others believe that our best hope lies in the creation of a new union of democracies. Plans also are underway, we have seen, for a world convention which will adopt a new constitution for the

COST OF NATIONAL DEFENSE
AND COST OF WORLD WAR I PER YEAR

$14.1 BILLIONS

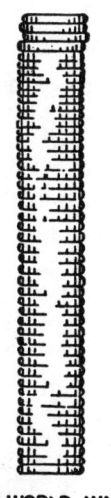

$10.7 BILLIONS

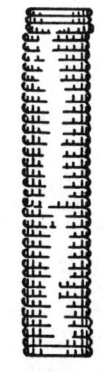

COST WORLD WAR I
(AVERAGE PER YEAR)

COST OF DEFENSE, 1947-48
AGAINST WORLD WAR III

(ANNUAL MONEY EXPENDITURES BY THE UNITED STATES ALONE)

VAST AMOUNTS OF MONEY FOR CONSTRUCTIVE PURPOSES WILL BECOME AVAILABLE IF THE PEACE CAN BE WON.

FIGURE 4[a]

entire world, or for as many nations as will meet the terms of membership.

Just now, the prospect for international cooperation and peace is dark. "Disintegrating forces are becoming evident," said Secretary Marshall in 1947. "The patient is sinking while the doctors deliberate."[1] The United States, discouraged by Russia's use of the veto and her general lack of cooperation, is by-passing the United Nations in respect to an increasing number of measures. Most important, the Marshall Plan provides that we act independently of the United Nations.

While among individuals and groups throughout the world there is a heartening spirit of generosity and friendliness, and while the nations of Western Europe are beginning to work together, we still appear to be placing greater faith in national defense through military power than in the power of good will and international cooperation. Shall this be our continued policy?

Plans proposed for national defense include a program of compulsory military service, which would probably cost between 2½ and 3½ billion dollars per year, a substantial increase in aircraft production, the arming and training of armies and navies of other American republics, and other cooperative measures with these nations, as well as extensive military research and development. These plans are covered only in part by a proposed budget for 1948-49 of over $11 billion and a supplementary budget of $3 billion. Finally, military authorities are now considering the decentralization of essential industries and the dwellings of the workers therein at a cost of billions of dollars, together with the utilization of underground caverns for industry capable of being sealed off against radioactivity and bacteria.[2]

Mighty Enterprise

This kind of total preparedness would require a high degree of totalitarianism. The large majority of citizens might not be able to choose their work and to live where they please. Conscription of labor would doubtless be necessary.[3]

The tragic aspect of this policy is the virtual certainty that it would lead to war. Yet defense seems necessary. We know that competitive armaments cannot avert war. But in the present state of international anarchy, superior military strength appears to be a reasonable *objective* for *each* nation. For the United States, armed strength surely seems to be a necessity until a workable world organization is created. A realist, anxious for a peaceful world, might say that for the present we must do two things—rely on force and at the same time seek to build a world organization strong enough to maintain peace. But can we do both?

While the nation's personal annual income in 1947 reached an all-time high of 197 billion dollars,[4] our national debt has become so great that the interest on it alone is over 5 billion dollars per year; and our programs for veterans cost us approximately 7½ billion dollars annually.[5] Can we afford to spend the many billions of dollars the government's program of national defense would necessitate, and at the same time appropriate additional billions required during the next few years by the Marshall Plan and similar measures to prevent the economic breakdown of Europe? Have we sufficient wealth to enable us to do both?

Furthermore, is there danger that if we invest our greater faith, energy, and wealth in military power we may find ourselves depending less and less on the United Nations or on some form of world government—the only *permanent* solution of the problem of war? "In the past," observes the Federation of American Scientists, "we have attained a measure of security by armed might. The atomic bomb and other methods of modern warfare have destroyed this possibility." Security, asserts the scientists, can be attained only through international cooperation.[6]

We have wealth, but we also have greater things than wealth. We excel in scientific research and invention. Politically, the people rule in the United States as in no other country. They elect the lawmakers and the executives they want. Notwithstanding all the weaknesses and dangers, reported earlier in this volume, the United States today, economically and politically, constitutes a great stronghold of democracy. Our influence is far-reaching. How shall we use this wealth, inventive genius, and influence?

The great issue now before us is this:

(1) Shall we continue to rely mainly on force, or (2) shall we invest our resources and our greater faith in good will and international cooperation—giving (a) generous economic aid to Europe and Asia, and (b) wholehearted leadership in the development of an international organization strong enough to maintain peace?

There can be only one answer. We know that in the end the only way to peace is through international organization. *Upon our willingness to give continuing aid and wholehearted leadership in the present crisis may depend the fate of western civilization.*

Modern war is the great anomaly. We destroy, then we rebuild—destroy and rebuild at the cost of billions upon billions of dollars sorely needed in man's warfare against his impersonal enemies—famine, poverty, crime and disease. We destroy, then we rehabilitate: in 1947 the United States, in its efforts to help save Europe, placed ally and enemy in

Mighty Enterprise

the same general category, giving aid not only to Britain and France but to Germany and Italy as well. If World War III suddenly bursts upon us, and if, conceivably, through some almost impossible stroke of good luck, we could destroy the enemy's economic power and armament in such a way as to prevent retaliation, *the logic of modern international economics would require us to participate at enormous expense in the rebuilding of the areas destroyed by the cataclysm.* Then why wait, logic inquires, for a task of rehabilitation more costly than the friendly aid to Europe (and even to Russia, if later the way becomes clear) that might prevent destructive conflict?

Furthermore, why should our government at Washington delay in exercising the leadership urgently needed to develop a stronger international organization—either a more effective United Nations or a world government? Such action would require little money. It might call for faith—great faith, perhaps. A strong international organization, with generous support from the United States, might increase the aid Europe and Asia now need after World War II, and it might prevent the almost inconceivable destruction of a third world war.

In 1790 the American people, under the leadership of Washington, Madison, and Hamilton, completed the organization of thirteen suspicious and antagonistic republics into a real union. The American people today have greater intelligence, greater resources, and greater influence for a more difficult and a more critical task—the mighty enterprise of creating a united, peaceful world.

What now can all of us, young and old, do to help build such a world, the one world which was our great objective a few years ago? Are not we, young and old, the people who create American opinion and policy? More than that, do we not as individuals and as a nation have an obligation to the rest of the world? Yes, we have surely arrived now, almost "without awareness, to a position of world leadership that we dare not fail to exercise."[7] But what can we do?

As a prerequisite to any action, we must seek understanding—understanding of the poverty and fear and frustrations of the people of Europe and Asia, particularly Russia; understanding of international relations; understanding (to the extent that that is now possible) of the reasons why men fight; understanding of the Marshall Plan in its relation to the future of Europe and the peace of the world; understanding of atomic energy. Especially important, we must inform ourselves regarding the larger implications of atomic energy. "The atomic scientists, I think," writes Dr. Albert Einstein, "have become convinced that they cannot arouse the American people to the truths of the atomic era by logic alone. There must be added that deep power of emotion which is a basic ingredient of religion. It is to be hoped that not only the churches but the schools, the colleges, and the leading organs of opinion will acquit themselves well of their unique responsibility in this regard."[8]

We must also inform ourselves regarding the United Nations and various proposals for world government. Then, on the basis of such understanding as may be gained, all of us, each for himself, must face the responsibility of deciding which plan of international organization in his opinion offers the greatest prospect of success. Thus we can make ourselves ready to render reasonably intelligent aid.

Specifically, what can we do?

First, as individuals and members of local communities, we can be friendly and generous, even at the cost of sacrifice. We can ship food to needy persons in Europe and Asia through CARE and similar agencies.* We can adopt a European town as did the people of Dunkirk, New York. Suppose that 1,000 of our cities and towns were to follow the example of that community of 20,000 people!

Each of us, too, can help remedy the present weaknesses in our democracy by joining a local, state, or national agency working for political, social, or economic reform. Local meetings can be organized, petitions circulated, letters sent to Congress, voters taken to the polls at election time. Those of us who have some influence in the field of industry can help avoid a great depression in "capitalistic United States," the signal for a great advance of communism. The people of Europe and Asia are watching us closely. They are reading the dispatches in the communist press regarding every indication that can be discovered of faults in our democracy. They want assurances that democracy is a better way than communism.

Second, each of us, individually and in groups, can work for the adoption of that plan for a united world believed to be most promising. The methods have been suggested —the distribution of books and pamphlets,† meetings, conferences, letters to congressmen, and other educational measures. Discussion groups may be especially helpful.‡ Consider the far-reaching effects of the efforts made by the young folks of Scarsdale, New York and of Oak Ridge, Tennessee, and by the older people of Baltimore, Mary-

* A list of these agencies can be found on p. 183.
† See list, pp. 185-189.
‡ See Questions for Discussion, p. 173.

land. A voluntary agency is available for the promotion of virtually every plan proposed for world organization; most of them may be joined by the individual citizen. So let us join the one of our preference and give it support—intelligent, generous support.

Will such measures do any good? Perhaps not. For we do not have full assurance that an "accident" in Europe will not lead speedily to World War III. John Fischer, however, thinks we may have a considerable period of stability resembling an armed truce, and that eventually— "perhaps in another generation—the truce may be converted into peace." If so, it "will be a long, tough, disagreeable job." But, with the aid of the young men and women who will constitute the next generation, the peace may be won.

"The free societies can generate forces for construction which will render impotent Soviet methods of destruction" asserts John Foster Dulles. "We must plan and act on a grand scale," he continues. "We must do mighty deeds such as are usually inspired only by war itself . . . The world will never have lasting peace so long as men reserve for war the finest human qualities. Peace, no less than war, requires idealism and self-sacrifice and a righteous and dynamic faith."[9]

The measures suggested, under the conditions outlined, are indispensable. For through such measures public opinion is formed, and in a democracy public opinion decides national policy. If you do not believe that, you do not believe in democracy. But more than that, the soundness of our country's choice, whether we decide to support the United Nations wholeheartedly or to help establish a new world government, *will depend in large measure upon the* extent to which the people inform themselves and the extent to which they participate in the discussion of the merits of the alternatives.

In a communist or fascist state the individual is trained to think to a pattern. Complete docility is expected. Indoctrination is surprisingly successful. In a democracy all peoples are free to do their own thinking. Statesmen alone cannot save us, or the generals, or the intellectuals. The participation of the people may provide the missing ingredient—if they bestir themselves.

The time is short. Momentous events are occurring with disturbing acceleration. Will it be survival or suicide?

We have suddenly been hurled into a new era with incalculable potentialities for good and evil.

We must think—fast and straight. And we must act!

A mighty enterprise awaits us.

Appendices

The inclusion of an organization in these appendices does not imply the endorsement by the editor of the policies and program of that organization, and the listing of a publication likewise does not indicate that the editor approves its contents. While agencies and publications frankly communistic will not be found in these lists, it is believed that, in general, the intelligent study of the problems of war and peace in their many ramifications requires a consideration of diverse points of view and schools of opinion.

Appendix A

Questions for Discussion

CHAPTER 1. *A New Era Ushers in a New and Greater Crisis*
1. Compare the good and the harm resulting from scientific discovery.
2. How can we develop the understanding and moral attitudes which will enable us more fully to utilize science for constructive purposes?

CHAPTER 2. *The Chaos of the Postwar World*
1. To what extent have we a responsibility for the welfare of the peoples of Europe and Asia?
2. What is the difference between fascism and communism? What is the best way to deal with communists in the United States?
3. How can we prevent another depression?
4. Why do we continue to mistreat those confined in county jails and mental hospitals? What can be done about it?

CHAPTER 3. *Russia, the Master Problem*
1. To what extent are the Russians justified in their fears?
2. To what extent does Russia's present attitude require the United States to build a stockpile of atom bombs?
3. Notwithstanding the Iron Curtain, how can the United States show greater friendliness towards Russia?

CHAPTER 4. *The Problem of National Defense*
1. What are your reasons for favoring or for disapproving

the measures of national defenses described in this chapter.
2. What are the best reasons for and against the adoption of universal military training?
3. Under what conditions may adequate defense require that industry "go underground?"

CHAPTER 5. *War in the Atomic Era*
1. What reasons may General Eisenhower have for believing that a next war will not last more than a few days or hours?
2. If we see another war approaching, to what extent must our government become a dictatorship?
3. What responsibility have you, as an individual, for the education of other people regarding the facts about war in the new era?

CHAPTER 6. *Getting Ready for World War III*
1. In view of the fact that poison gas was not used in World War II, do you think that atomic radiation and biological warfare will be used in World War III, if it comes? Please discuss.
2. In your opinion, how would a third world war be started?
3. To what extent are Dr. Urey's fears justified?
4. Was the United States justified in making the atomic bomb? In using it? What are your reasons for thinking so?

CHAPTER 7. *The Need for Understanding*
1. How can young and old be induced to give more attention to international affairs in preference to relatively unimportant matters?
2. Should the United States assume the world leadership proposed in this chapter? If so, in what fields and in what ways?
3. What measures, if any, should be taken to prepare the people of this country for world leadership?

CHAPTER 8. *The Kind of Understanding Needed*
1. Do you think more students should be induced to study the social sciences in preference to the natural sciences? If so, how?
2. To what extent is the author's concept of patriotism a valid one?
3. How can all science be unified as suggested?

CHAPTER 9. *War and Human Nature*
1. Of the causes of war suggested in this chapter which do you think are the most real—the warlike nature of man, social

Appendices

injustice, frustration, the need for territory and markets, rivalries for power?
2. Which of the substitutes suggested for war do you consider the most promising? Why do you think so?
3. To what extent, if any, do you believe that psychologists and other social scientists should participate in the work of the United Nations and its subsidiary bodies?
4. If you had wealth, what proportion of it, if any, would you give to psychologists for further research?

CHAPTER 10. *The Problem of Understanding Russia*
1. Since our great navy and our stockpile of bombs cause other nations to fear us, to what extent, if any, should we reduce our present armaments?
2. To what extent, if any, shall the United States take the leadership in establishing such "Authorities" in Palestine and elsewhere as suggested in this chapter?
3. How can we help convert a military truce into real international peace?

CHAPTER 11. *The Larger Implications of Atomic Energy*
1. Which of the constructive uses of atomic energy do you think are the most promising?
2. How can we be sure that the development of more mechanical power will result in greater social justice and improved living conditions for all?
3. How great is the need for scientific students devoted to the constructive use of science?

CHAPTER 12. *Training and Understanding for Youth*
1. How would you answer the criticism that in some colleges social science courses are "easy" and attract the less competent students?
2. What extracurricular activities in school or in college appeal to you as the most educative?
3. What suggestions regarding vacation activities appeal to you as having greatest educational value?

CHAPTER 13. *The United Nations*
1. How can our representatives in the United Nations persuade Russia to be more cooperative?
2. In what other ways do you think the United Nations can be more effective?
3. What specialized agency do you think is doing the most important work and why do you think so?

CHAPTER 14. *Proposals for World Government*
1. To what extent is it true that the international situation now is similar to the situation among the colonies after the Revolution?
2. Do you think that the United Nations without Charter amendments can maintain the peace or do you think that a world government is necessary? What are your reasons?
3. If it appears that world government is necessary, which plan would you prefer and why?
4. Discuss the issue between Dr. Urey on the one hand and Drs. Morrison and Wilson on the other hand.

CHAPTER 15. *Local Groups in Action*
1. What kind of local action might be most effective educationally in your school, college, or local community?
2. In your opinion, to what extent should an educational program emphasize the scientific, and to what extent the political aspects of the present crisis?

CHAPTER 16. *Beginnings of World-wide Friendliness*
1. Which of the various community programs described do you think are the most promising, and why?
2. How can an individual most effectively aid European and Asiatic peoples?

CHAPTER 17. *The Marshall Plan—An Adventure in Friendliness*
1. Might it have been possible to include Russia in the European Recovery Program. If so, through what means?
2. Discuss the far-reaching benefits of the program for which our leaders hope.

CHAPTER 18. *Force and Friendliness*
1. How serious, in your judgment, are the dangers of an "accident" and present trends toward war?
2. Do you believe the United States should exercise the greater faith and friendliness suggested? If so, specifically what plans would you approve?
3. Do you think the proposed board of strategy is desirable and practicable? If so, how would you set it up and develop it?

CHAPTER 19. *The Mighty Enterprise*
1. Do you think that an armed truce may continue for twenty years or more or do you think that an "accident" may occur to precipitate war?

Appendices

2. Just how true is the statement in this chapter: "The soundness of our country's choice, whether we decide to support the United Nations wholeheartedly or to help establish a new world government, *will depend in large measure upon the extent to which they* [the people] *participate in the discussion of the merits of the alternatives.*"?
3. If an armed truce continues until another generation, do you think that the next generation will be better prepared to deal with the crisis? Please discuss.
4. Discuss in general the relative merits of various major and minor ways to peace set forth in this book.

Appendix B

Voluntary Organizations Promoting World Peace

1. General educational agencies

(a) New agencies on atomic warfare and international organization

The American Association for the United Nations, Inc. (45 E. 65th St., New York 21, New York) and its research affiliate, the *Commission to Study the Organization of Peace*, are devoted solely to vigorous and militant educational programs on behalf of the United Nations. Its youth affiliates are the "Collegiate Council for the United Nations," for college students, and "United Nations Youth" for high school students. The director of the Association is Clark M. Eichelberger. Minimum membership fee is $2.00.

The United World Federalists (31 E. 74th St., New York 21, New York) is seeking the revision of the United Nations Charter to provide world government. UWF is the result of a merger of five world government membership organizations on February 22, 1947, at Asheville, North Carolina. Membership is open, and the dues are $3.00 per year. The UWF has almost 300 chapters in 37 states, with over 16,000 members.

The student division of UWF, of which Helen Ball is the director, has about 150 chapters in the colleges and high schools of the country. Membership in the division is open to students and others. Dues are $1.00 per year.

Federal Union, Inc. (700 9th St., N.W., Washington, D.C.) advocates a union now of existing democracies. Clarence K. Streit is the president.

Appendices 179

The Citizens Committee for United Nations Reform, Inc. (16a 62nd St., New York 21, New York) is the committee promoting the "Quota Force Plan." Ely Culbertson is the acting chairman. Membership is open to all interested persons. There are no dues.

World Republic (35 E. Wacker Drive, Chicago, Illinois) promotes the establishment of world government through a world constitutional convention. Dues are $1.00 per year. World Republic was started by seven students with a one-car garage for an office, and has expanded to an organization with a membership of over 5,000. The national chairman is Rempfer L. Whitehouse.

The Campaign for World Government (343 S. Dearborn St., Chicago) is an educational agency working for a federation of nations open at all times to all nations. It believes law should be enforced through federal civilian police action on individual citizens. The executive secretary is Georgia Lloyd.

Common Cause (22 E. 60th St., New York 19, New York) has been launched to serve all who wish to take part "in the great rebirth of democratic faith in America and the world." The chairman is Mrs. Wales Latham. Membership is open to all interested.

The Committee to Frame a World Constitution (975 E. 60th St., Chicago 37, Illinois) was organized at the University of Chicago in 1945 to develop a blueprint for world government which would present for consideration and discussion the specific plans and mechanics of a world state at a world constitutional convention about 1950. The president of the Committee is Robert M. Hutchins, Chancellor of the University. A preliminary draft of a world constitution was completed early in 1948.

Women's Action Committee for Lasting Peace is organized to unite American women to work for full participation by the United States in the United Nations. General members pay $1.00 a year, and sponsors $10.00 a year. All members receive regularly a monthly bulletin and other material.

American Veterans Committee (1860 Broadway, New York). The educational program of this organization gives special attention to international affairs.

World Movement for Federal World Government. M. Jean Larmeroux is president. His address is Comité International des Etate Unis du Monde, 12 Avenue George V, Paris, France.

The Youth Council on the Atomic Crisis (High School, Oak Ridge, Tenn.). This is the central organization which provides speakers for conferences and meetings. There are councils in various states. The faculty adviser is Philip E. Kennedy.

The Federation of American Scientists, (1749 L St., N. W., Washington 6, D.C.) is an organization, located at the nation's capital, with 19 member associations of scientists. It is especially interested in legislation pertaining to atomic energy. The chairman is Robert E. Marshak.

The Emergency Committee of Atomic Scientists (625 Madison Avenue, N.Y.) is composed of a small group of scientists, headed by Dr. Albert Einstein, who are doing their utmost to bring home to the people of the United States the simple facts of atomic energy and their tremendous implications for world peace or world destruction. By conference and letter these men have reached directly over a quarter of a million key citizens. The committee is seeking to raise a minimum of $1,000,000 and has already received a substantial part of this sum.

The Association of Scientists for Atomic Education (625 Madison Avenue, N.Y.) is an organization of about 800 physical scientists most of whom worked on the atomic energy project. Members have given over 1000 talks; literature on atomic energy has been sent to more than 2,000 foreign scientists; talks have been broadcast over local stations and major networks. Conferences have been conducted in about 50 local communities; high schools have been advised on curricular revisions in respect to atomic energy developments.

The National Committee on Atomic Information (1749 L St., N.W., Washington, D.C.) consists of 60 member agencies. It carries out an extensive educational program. Its director is George L. Glasheen.

The Atomic Scientists of Chicago (1126 E. 59th St., Chicago 37, Illinois). While this group is only one of 19 local and state organizations of atomic scientists, it is included here because it publishes once a month the *Bulletin of Atomic Scientists*. The cost is $2.00 per year. This periodical contains articles and reports from the United Nations Atomic Energy Commission and the Association of Scientists for Atomic Education.

Appendices

(b) Longer Established Agencies

American Friends Service Committee, 20 S. 12th St., Philadelphia
American Institute of Pacific Relations, 1 E. 54th St., New York
American Youth for a Free World, 144 Bleecker St., New York
Americans for Democratic Action, 9 E. 46th St., New York
Carnegie Endowment for International Peace, 405 W. 117th St., New York
Church Peace Union, 70 Fifth Ave., New York
Council on Foreign Relations, 58 E. 68th St., New York
Federal Council of Churches, Department of International Justice and Good Will, 105 E. 22nd St., New York
Fellowship of Reconciliation, 2929 Broadway, New York
Foreign Policy Association, 8 W. 40th St., New York
Friends of Democracy, Inc., 137 E. 57th St., New York
Institute of International Education, 2 W. 45th St., New York
National Catholic Welfare Conference, 1312 Massachusetts Ave., Washington, D.C.
National Council for the Prevention of War, 1013 18th St., N.W., Washington 6, D.C.
Royal Institute of International Affairs, 542 Fifth Ave., New York
Students International Union, Inc., 522 Fifth Ave., New York
Women's International League for Peace and Freedom; 532 17th St., Washington, D.C.
Woodrow Wilson Foundation, 45 E. 65th St., New York
World Alliance for International Friendship Through the Churches, 70 Fifth Avenue, New York
World Education Service Council, 2 W. 45th St., New York
World Peace Foundation, 40 Mt. Vernon St., Boston, Massachusetts

2. *Agencies promoting institutes on international relations, and international visitations for study and service*

American Friends Service Committee, 20 S. 12th St., Philadelphia, conducts the following summer projects for students—
  Peace Caravans—Students in caravans travel over an area rousing interest and concern in the problems of world peace.
  Institutes of International Relations—Ten days' discussion of international affairs held in various areas of the United States.

Mexican Service Seminars—Students work with Mexican people on public health and education in the states of Morelas and Hidalgo in Mexico.

International Service Seminars—Representatives of many different nations and cultures who are studying in the United States are brought together in various areas for seven weeks of cooperative living and studying.

The Experiment in International Living, Putney, Vermont.

This organization takes young Americans to live as members of families of education and culture in other countries. Its activities are described in a booklet *Crossroads*, free on request.

Department of State, Division of International Exchange of Persons, Student and Trainee Branch, Washington, D.C.

Grants for Study in other American Republics—The Department of State has resumed on a limited basis the program of travel and maintenance grants to assist United States graduate students to undertake studies and research in other American republics.

Fulbright Act—Programs of student exchange. Financial assistance may be provided for United States citizens studying in foreign countries and for citizens of those countries for study and travel in the United States. Under the Smith-Mundt Act, similar assistance may be secured.

Public Law No. 346, the "G. I. Bill of Rights"—Requests for information on study abroad under this Bill should be addressed to the Veterans Administration, Washington, D.C.

UNESCO, United Nations Educational, Scientific and Cultural Organization, Lake Success, New York.

This organization is setting up scholarships on an exchange basis among nations.

(*Organizations assisting advanced students to study abroad in particular areas or fields*)

Institute of International Education, 2 W. 45th St., New York

The Rockefeller Foundation, 49 W. 49th St., New York—Public health, medical sciences, natural science, social sciences, humanities.

The Commonwealth Fund, 41 E. 57th St., New York—Education, health, medical education and research, mental hygiene.

American Council of Learned Societies, 1219 16th St., N.W., Washington, D.C.—Humanistic and social studies.

Social Science Research Council, 230 Park Ave., New York—Social sciences.

Appendices 183

3. Agencies through which money and other aid
may be sent to foreign peoples

American Friends Service Committee, 20 S. 12th St., Philadelphia
American Jewish Joint Distribution Committee, 270 Madison Ave., New York
CARE, Cooperative Associations for Remittances to Europe, 50 Broad Street, New York
Church World Service, 37 E. 36th St., New York
International Committee of Y.M.C.A., 347 Madison Ave., New York
National Board of Y.W.C.A., 600 Lexington Ave., New York
United Jewish Appeal, Inc., 342 Madison Ave., New York
War Relief Service—National Catholic Welfare Conference, 350 Fifth Avenue, New York
World Student Service Fund, 20 W. 40th St., New York

(For further information and lists of denominational agencies, of agencies serving specific countries in Europe and Asia, and of specialized agencies on child welfare, displaced persons, and comparable problems, the following organizations may be consulted.)

Advisory Committee on Voluntary Foreign Aid of the United States Government, Washington 25, D.C.
American Council of Voluntary Agencies for Foreign Service, 122 22nd St., New York

4. Agencies through which the exchange of letters may be developed

American Junior Red Cross (National Headquarters, 17th and D Sts., N.W., Washington, D.C.)
This organization promotes a group correspondence plan by means of letter-booklets. Any Junior Red Cross group through senior high school may correspond with similar groups in any other country where Junior Red Cross societies exist. National headquarters will send a letter-booklet to the Red Cross society in a selected country. There it will be translated and sent to a suitable school for exchange. For membership in Junior Red Cross, high schools and preparatory schools pay $1.00 for a group of 100 or less.

Division of International Educational Relations (U. S. Office of Education, Washington 25, D.C.)

This Division attempts to handle correspondence between individual students, teachers, classes, and schools. It will make every effort to find suitable correspondents from the many foreign requests received. The majority of requests are from Germany and Austria—not only from students of all ages, but business and professional people as well. Write to Dr. Helen Dwight Reid, Division of International Educational Relations, address above.

International Friendship League (41 Mt. Vernon St., Beacon Hill, Boston, Massachusetts)

This league collects names, ages, addresses, ocupations of parents, and special hobbies of students in many countries. American students pay a life membership of $.50. Registration card must be signed by a teacher or other qualified adult. The league matches ages and special interest, and sends from three to six names to each member. Write to Miss Edna MacDonough, executive secretary.

Appendix C

Educational Materials on War and Ways to Peace

Books

James F. Byrnes: *Speaking Frankly*, Harper & Brothers, New York, $3.50
This is former Secretary Byrnes' "off-the-record" story of the two years after V-E Day. It is a fascinating book of historical significance.

Ansley J. Coale: *The Problem of Reducing Vulnerability to Atomic Bombs*, Princeton University Press, Princeton, New Jersey, $2.00.
Written under the direction of a notable group of social and physical scientists, this is a precise and straightforward statement analyzing problems of atomic energy. An important book of special interest to college students interested in technical problems of defense.

Norman Cousins: *Modern Man is Obsolete*, The Viking Press, New York, $1.00.
An essay written the night following the news of Hiroshima. Mr. Cousins challenges and stimulates the reader to think for himself on the vast issues confronting modern man.

Ely Culbertson: *Total Peace*, Doubleday, Doran & Co., New York, $2.50.
A controversial book on plans for winning the peace through collective defense and an international police force. It also presents new principles of United States foreign policy.

Eidinoff and Ruchlis: *Atomics for the Millions*, Whittlesey House, New York, $3.50.
Introduction by Dr. Harold C. Urey. The Chicago *Tribune* says:

"Of all the atom books this does one of the best jobs in substituting human interest and intelligible language for the jargon of the laboratory."

John Fischer: *Why They Behave Like Russians*, Harper & Brothers, New York, $2.75.

This book, although written with informality and humor, presents a sincere, fair, and objective picture of life and conditions in Russia today.

Thomas Galt: *How the United Nations Works*, Crowell Co., New York, $2.00.

Of special interest to high school students. A history of the development of the United Nations and an analysis of its functioning illustrated with amusing and lively cartoons. Exceedingly valuable for its factual data and as a reference book for schools.

John R. Hersey: *Hiroshima*, Alfred A. Knopf, Inc., New York, $1.75.

A story about the atomic bombing of Hiroshima as it affected six persons. Straightforward, clear, objective reporting, it is considered one of the classics of the war.

Julia E. Johnsen: *The Atomic Bomb*, The Reference Shelf, vol. 19, no. 2, The H. W. Wilson Co., New York, $1.25.

This book contains a collection of articles by outstanding scientists, statesmen, and other writers presenting the factual background and other helpful material concerning atomic fission, its principles and controls.

Cord Meyer, Jr.: *Peace or Anarchy*, Little, Brown & Co., New York, $2.50.

A leading spokesman of world government analyzes the world situation and states how to give the United Nations the permanent peace-keeping authority it needs. A valuable and disturbing book.

Gardner Murphy, Editor: *Human Nature and Enduring Peace*, Houghton Mifflin Co., New York, $4.00

This is the Third Yearbook of the Society for the Psychological Study of Social Issues. Despite its technical aspect this is a very readable book stressing the many facets of the problem of human nature as it relates to war. The contributions from fifty different specialists represent many viewpoints.

Emery Reves: *The Anatomy of Peace*, Harper & Brothers, New York, $2.00.

Mr. Reves presents his case for world government and analyzes the causes of war. Especially provocative and stimulating for students of international affairs.

Appendices

Leland Stowe: *While Time Remains*, Alfred A. Knopf, Inc., New York, $3.50.
A correspondent who has literally seen history in the making, has written this book with insight, passion, and humanity. He discusses our revolutionary world, the new Europe, and the new Soviet power. A challenging and stimulating appraisal of the current world.

Raymond Swing: *In the Name of Sanity*, Harper & Brothers, New York, $1.00.
This book, based on a series of broadcasts, is especially recommended by Dr. Einstein.

Wendell L. Willkie: *One World*, Simon and Schuster, Inc., New York, $1.00.
Following his 49-day world tour in 1942, Mr. Willkie wrote this plea for unity and understanding among all peoples of the world. It is a simple, readable report stressing the nearness and interdependence of the peoples of all continents.

Harris Wofford: *It's Up To Us*, Harcourt, Brace and Co., New York, $2.00.
The founder of the Student Federalist movement presents the story of the beginnings and aims of that organization. It is one American youth's answer to the problems facing the United Nations. It records with freshness, zest, and sincerity how some younger Americans are thinking on world affairs. Of interest to young and old.

The U. N. Yearbook, foreword by Secretary General Trygve Lie, Columbia University Press, New York, $10.00.
This is an important source book of United Nations accomplishments from the beginning of the organization to July 1, 1947. Its 991 pages contain background history, a full description of UN workings, an appendix of important documents, and statistical details of the member states, as well as biographical information on UN delegates and officials.

Pamphlets

The Atomic Bomb vs. Civilization, Robert M. Hutchins, Human Events, Inc., Henry Regnery Co., 53 S. Washington St., Hinsdale, Illinois, $.20.

The Atom Bomb, a World Problem, Vernon H. Holloway, 289 Fourth Ave., New York, $.25.

Atomic Energy in the Coming Era, David Dietz, Dodd, Mead & Co., $.25.

The Atomic Age Opens, Wendt and Geddes, The World Publishing Co., $.25.

This Atomic World, Board of Christian Education of the Presbyterian Church, 830 Witherspoon Building, Philadelphia, $.10.

Federal Union of the Free, Clarence K. Streit, Federal Union, Inc., 700 Ninth St., N. W., Washington, D.C., $.10, 20 for $1.00.

A Guide for Speakers and Teachers to the United Nations, (United Nations Publication) International Documents of the Columbia University Press, New York, $.50.

I'm a Frightened Man, Harold C. Urey, Committee on Atomic Information, 1749 L St., N.W., Washington, D.C., $.10, 25 for $1.00.

The International Control of Atomic Energy, a compendium of articles published in the Bulletin of the Atomic Scientists. Available from Bulletin of the Atomic Scientists, 1126 E. 59th Street, Chicago 37, Illinois.

Structure of the United Nations, (a United Nations Publication) Columbia University Press, New York, $.35.

United Nations in Action, special issue (October 20, 1947) of Scholastic Magazine, 220 E. 42nd St., New York, $.35.

We the Peoples—A Brief History of the United Nations, 1947 revised edition. Mrs. Harrison Thomas, American Association for the United Nations, Inc., 45 E. 65th St., New York, $.20.

Your Flesh Should Creep, Joseph and Stewart Alsop, Committee on Atomic Information, 1749 L St., N.W., Washington, D.C., free on request.

Sources from which Pamphlets can be Obtained

* American Association for the United Nations, Inc., 45 E. 65th St., New York

American Friends Service Committee, 20 S. 12th St., Philadelphia

American Library Association, Exploring the Times Series, 520 N. Michigan Ave., Chicago

Carnegie Endowment for International Peace, 405 W. 117th St., New York

Department of State, Washington, D.C.

Federal Council of Churches, Department of International Justice and Good Will, 105 E. 22nd St., New York

* Foreign Policy Association, Headline Series, 22 E. 38th St., New York, $.35 each

Forerunners Publications, 2929 Broadway, New York

Henry Regnery Co., Human Events Pamphlets, 53 S. Washington St., Hinsdale, Illinois

International Documents Service, Columbia University Press, New York, Official Publications of the United Nations

National Catholic Welfare Conference, 1312 Massachusetts Ave., Washington, D.C.

* Public Affairs Pamphlets, 22 E. 38th St., New York, $.20 each

* Town Meeting Reports, Town Hall, Inc., 123 E. 43rd St., New York, $.10 each

United Nations Department of Public Information, Lake Success, New York

University of Chicago Round Table, Chicago, Illinois, $.10 each

Woodrow Wilson Foundation, 45 E. 65th St., New York

Sources from which Films Can Be Obtained

Association Films, 347 Madison Ave., New York, or 19 La Salle St., Chicago

Brandon Films, 1600 Broadway, New York

Julien Bryan Distributors, International Film Foundation, Educational Film Library Association, 1600 Broadway, New York

Carnegie Endowment for International Peace, 405 W. 117th St., New York

Encyclopedia Britannica Films, 1841 Broadway, New York

Film Publishers, Inc., 25 Broad St., New York

Film Section of the United Nations Educational, Scientific and Cultural Organization, Lake Success, New York

March of Time, Inc., 369 Lexington Ave., New York

The Motion Picture Producers Association, 28 W. 44th St., New York

* Each of these organizations publishes various pamphlets of special value to the general reader. Write to them for lists.

Protestant Film Commission, 45 Astor Place, New York
United Nations Department of Public Information, Films and Visual Information Division, Lake Success, New York. Send for catalogue "The United Nations in Films"
United World Films, Inc., 445 Park Ave., New York

Periodicals

American Observer, 1733 K St., N.W., Washington 6, D.C., weekly, $2.00 per year

Atomic Information, 1749 L St., N.W., Washington, D.C., monthly, $2.00 per year

Bulletin of the Atomic Scientists, 1126 E. 59th St., Chicago 37, Illinois, monthly, $.25 per copy, $2.50 per year

Changing World, American Association for the United Nations, 45 E. 65th St., New York 21, New York, monthly, $.10 per copy, $1.00 per year

Foreign Affairs Outlines on Building the Peace, Department of State, Washington, D.C., quarterly, no charge

Foreign Affairs, an American Quarterly Review, Council on Foreign Relations, Inc., 58 E. 68th St., New York, $5.00 per year

Freedom and Union, journal of the World Republic, 700 9th St., N.W., Washington 1, D.C., monthly, $.35 per copy, $4.00 per year

International Conciliation, made up largely of official documents, 405 W. 117th St., New York 27, New York, monthly, $.25 per year, $1.00 for five years

Senior Scholastic, a magazine for high school students, 200 E. 42nd St., New York 17, New York, weekly during the school year, $.10 per copy, $1.50 per year

Science Illustrated, contains considerable material on atomic energy, McGraw-Hill Publishing Company, Broadway at 11th St., Louisville, Kentucky, monthly, $.25 per copy, $3.00 per year

The Student Federalist, 31 E. 74th St., New York 27, New York, monthly, $1.00 per year

United Nations News, Woodrow Wilson Foundation, 45 E. 65th St., New York 21, New York, monthly, $3.00 per year

United Nations Bulletin, Lake Success, New York, semi-monthly, $3.00 per year

United Nations World, The John Day Company, 385 Madison Ave., New York 17, New York, $.35 per copy, $4.00 per year

World Government News, 215 Third Ave., New York 3, New York, monthly, $.10 per copy, $1.00 per year

U.S. News—World Report, 435 Parker Ave., Dayton, Ohio, weekly, $.15 per copy, $5.00 per year

References

References

CHAPTER 2

1. From conversation with Hanson W. Baldwin's office, New York Times, February, 1948; Clark M. Eichelberger: editorial, Changing World, January, 1948, p. 3.
2. "The Truth about Russia's 12,000,000 Slave Laborers," Look, October 28, 1947, p. 60; Time, September 1, 1947, p. 84; New York Times, December 11, 1947, p. 32; Milton Lipson: "Terror: The World's Fastest Growing Business," The United Nations World, February, 1948, pp. 32-35.
3. The Senior Scholastic, October 20, 1947, p. 40.
4. New York Times, February 13, 1948, editorial, "Free to Starve?" see also p. 109.
5. Chester Bowles: "To Bring Hope To Children Without Hope," New York Times Magazine, February 1, 1948, pp. 10, 11, 35.
6. Beulah Amidon: "Hunger," Survey Graphic, July, 1947, p. 373.
7. New York Times, August 20, p. 16 and August 25, 1947, p. 5.
8. New York Times, February 5, 1948, pp. 1, 10.
9. New York Herald Tribune, February 17, 1948.
10. Foster Hailey: "Too Many People on Too Little Land," New York Times Magazine, March 2, 1947, p. 5.
11. Gertrude Samuels: "The Unheard Cry of the World's Children," New York Times, October 12, 1947, p. 58; New York Times, February 2, 1948, p. 7.
12. Time, June 2, 1947, p. 31; see also New York Times Magazine, October 26, 1947.
13. Christian Century, February 25, 1948, pp. 228-9.
14. New York Times, February 5, 1948, p. 22.
15. Leland Stowe: While Time Remains, Alfred A. Knopf, New York, Ch. X.
16. Mallory Browne, New York Times, June 15, 1947, p. 1.
17. Ellis Arnall: "My Battle Against the Klan," Coronet, October, 1946, pp. 3-8.
18. Ernest O. Hauser: "The True Meaning of the Iron Curtain," Saturday Evening Post, June 14, 1947, p. 158.

19. New York Herald Tribune, September 5, 1947, p. 12.
20. New York Daily News, July 30, 1947, p. 41.
21. Time, April 21, 1947, p. 31; Look, April 15, 1947, p. 32, et seq.; New York Herald Tribune, April 14, 1947, p. 21.
22. Life, May 6, 1946, p. 102.
23. New York Herald Tribune, August 29, 1947, p. 10.
24. Christian Century, October 29, 1947, p. 1291.
25. E. Stanley Jones: "India's Caste System and Ours," Christian Century, August 20, 1947, p. 995.
26. Newsweek, February 10, 1947, p. 25.

CHAPTER 4

1. Chart "Revenues and Expenditures, Federal Government, Fiscal year ending June 30, 1948 (Revised August, 1947)" $10.4 billions plus $320 millions for Greek-Turkish aid (Letter from Donald B. MacPhail, Bureau of the Budget, October 17, 1947) gives a total of $10.7 billions.
2. New York Times, February 27, 1948, p. 1.
3. New York Herald Tribune, April 3, 1948, pp. 1, 24.
4. Letter from the Book and Magazine Section, Office of Public Information, Navy Department, Washington, November 25, 1947.
5. New York Times, June 2, 1947, pp. 1, 5.
6. Alexander Stewart: "Is 'Umtee' the Answer?" Christian Century, May 28, 1947, pp. 680, 681; also letter of July 1, 1947 from Hanson W. Baldwin of the New York Times.
7. New York Times, April 1, 1948, p. 10.
8. "Survival in the Air Age." Report of the President's Air Policy Commission, Washington, January 1, 1948, p. 25.
9. Thomas K. Finletter: "Air Power and World Peace," The Atlantic, April, 1948, pp. 25, 26.
10. New York Times, January 13, 1948, p. 17; New York Times, April 2, 1948, p. 19.
11. Daily News, New York, June 24, 1947, p. 6.
12. New York Times, June 26, 1947, p. 1.
13. Christian Century, June 11, 1947, p. 732.
14. New York Times, January 13, 1948, pp. 16, 20.
15. New York Times, January 13, 1948, p. 17.
16. New York Times, April 2, 1948, p. 19.
17. New York Times, March 31, 1948, p. 18.
18. J. Marschak, E. Teller, and L. R. Klein: "Dispersal of Cities and Industries," Bulletin of the Atomic Scientists, Vol. 1, No. 9, p. 15.
19. Cord Meyer, Jr.: "What Price Preparedness?" Atlantic Monthly, June, 1947, p. 30.
20. James Burnham, The Struggle for the World, John Day, New York, 1947.
21. "Blueprint for Empire," Christian Century, May 21, 1947, pp. 646, 648.
22. New York Times, February 11, 1948, p. 11.
23. James Burnham, op. cit., pp. 135, 183.
24. James Burnham, op. cit., pp. 242, 232.

References

Chapter 5

1. A one-page statement, not dated, by the Emergency Committee of Atomic Scientists, 90 Nassau Street, Princeton, New Jersey.
2. *The Atomic Bomb—Facts and Implication*, a pamphlet published by the Atomic Scientists of Chicago, 1946.
3. Irving Langmuir: "An Atomic Arms Race and its Alternatives," from *One World or None*, McGraw-Hill Book Company, 1946, p. 47; *Time*, February 24, 1947, p. 96.
4. R. Magruder Dobie: "How We'll Travel Faster than Sound," *This Week* (New York Herald Tribune) November 2, 1947, p. 7.
5. *Time*, June 23, 1947, p. 70.
6. Milton Silverman: "Operation Hotfoot," *Saturday Evening Post*, December 14, 1946, p. 20, et seq.
7. Ansley J. Coale: *The Problem of Reducing Vulnerability to Atomic Bombs*, Princeton University Press, 1947, pp. 69, 76, 77.
8. David B. Parker: "Mist of Death over New York," *Reader's Digest*, July, 1947, pp. 7-10, condensed from the *Coast Artillery Journal*, March-April, 1947.
9. Ladislas Farago: "Tabun," *United Nations World*, February, 1948, pp. 11-13.
10. Kenneth V. Thimann: "The Role of Biologists in Warfare," *Bulletin of the Atomic Scientists*, August, 1947, pp. 211, 212; and letters from Dr. Thimann, November 13 and 17, 1947.
11. Gerald Wendt: "Silent Death," *Science Illustrated*, October, 1946, pp. 25, et seq.
12. Robert S. Allen's column in *Liberty*, July 19, 1947, p. 14.
13. Encyclopedia Americana Vol. 28, 1946, p. 650; World Almanac 1947, p. 523.
14. Letters from Donald B. McPhail, Bureau of the Budget, Washington, D.C., August 27, September 10, October 17, 1947, and accompanying charts.
15. "What is a Billion?" pictorial article, Washington *Daily News*, May 29, 1946.
16. Based on a cost of $336,000,000,000 for 1,345 days.
17. Memorandum of Columbia University Statistics as of June 30, 1946, p. 2; letter of May 13, 1947 from the Commandant, Third Naval District, 90 Church Street, New York 7, New York.
18. Quincy Wright: *A Study of War*, University of Chicago Press, 1942, p. 675; *Enginemen's Magazine*, September, 1944.
19. Metropolitan Life Insurance Company Statistical Bulletin, January, 1946, p. 7; *World Almanac*, 1947, p. 521; letter from the editor, *World Almanac*, June 10, 1947. (22,060,000 total dead less 10,000,000 military dead leaves about 12,000,000 dead civilians).
20. Nat S. Finney: "We Can Lose the Next War in Seven Days," *Look*, July 8, 1947, p. 21.

Chapter 6

1. New York *Times*, September 6, 1946, p. 16.
2. *Time*, March 3, 1947, p. 67.
3. *Time*, September 2, 1946, p. 52.
4. Freyling Foster in *Collier's*, August 30, 1947, p. 6.
5. New York *Times*, August 4, 1947, p 1.
6. Joseph and Stewart Alsop: "Dragons are Old Fashioned," New York *Herald Tribune*, July 11, 1947; p. 17.
7. New York *Times*, October 1, 1947, p. 27.
8. New York *Times*, March 8, 1947, p. 4.

Chapter 10

1. *Christian Century*, August 20, 1947, p. 1006.

Chapter 11

1. This paragraph is based on a memorandum from F. R. Trapnell, U.S. Atomic Energy Commission, Washington, D.C.
2. Ibid.
3. John W. Campbell: *The Atomic Story*, Henry Holt and Company, Inc., New York, pp. 266-268, 274.
4. Op. cit., p. 263.
5. Cuthbert Daniel and Arthur M. Squires, *Bulletin of the Atomic Scientists*, April-May, 1947, p. 111; David F. Cavers, op. cit., October, 1947, p. 283; New York *Times*, April 5, 1948, p. 1. The footnote, except the last paragraph, was formulated with the assistance of Dr. Squires, who approved it February 26, 1948.

Note. Members of the technical staff of the United States Atomic Energy Commission reviewed the material covered by the references below. In a few instances revisions were made. (Commission's letter of November 21, 1947). Thus, in these instances data in the text are somewhat different from the data in the sources below.

6. "A Report of the International Control of Atomic Energy," prepared for the Secretary of State's Committee on Atomic Energy, Department of State Publication 2498, March 16, 1946, p. 19.
7. *Newsweek*, August 11, 1947, p. 55.
8. *Newsweek*, op. cit., pp. 55, 56.
9. *Time*, August 11, 1947, p. 74.
10. *Newsweek*, op. cit., pp. 55-58.
11. New York *Times*, August 12, 1947, p. 1.
12. New York *Times*, August 21, 1947, p. 8.
13. *Newsweek*, September 1, 1947.
14. William L. Laurence: Dispatch from St. Louis, New York *Times*, September 4, 1947, p. 1.
15. John W. Campbell, op. cit., p. 279.

References

Chapter 12

1. Ernie Pyle: *Brave Men*, Henry Holt & Company, 1944, p. 319.
2. *Life*, June 23, 1947, p. 93; an 8-page, undated circular from the State Department: "The Foreign Service of the United States."
3. Edith Iglauer: "Housekeeping for the Family of Nations," *Reader's Digest*, August, 1947, pp. 105-108, condensed from *Harper's Magazine*, April, 1947; Report of the Preparatory Commission of the United Nations, published for the United Nations by His Majesty's Stationery Office, London, 1946, Chap. VIII.
4. Benjamin Fine: "Education in Review," New York *Times*, May 25, 1947, Sec. IV, p. 9.
5. Letter from Sydney Holmes, New York *Times*, June 18, 1947.
6. *Herald Statesman*, Yonkers, New York, April 28, 1947, p. 1; *Time*, May 12, 1947, pp. 68, 69.
7. New York *Times*, July 6, 1947, Sec. IV, p. 7.
8. *Changing World*, published by the American Association for the United Nations, June, 1947, p. 3.
9. *Christian Century*, May 14, 1947, p. 640.
10. New York *Times*, July 10, p. 5, and July 11, p. 5, 1947; also information telephoned from U.N. Training Division, February 27, 1948.
11. New York *Times*, June 30, p. 21, and July 1, 1947. Revised by the Institute of International Education, Nov. 20, 1947.
12. New York *Times*, July 1, 1947, p. 27.
13. Based on a memorandum, Nov. 20, 1947, from the Division of International Exchange of Persons, Dept. of State, Washington, D.C.

Chapter 13

1. Releases from the ICEF, January and February 10, 1948; letter from Gilbert Redfern, ICEF, New York, February 3, 1948; New York *Times*, January 12, 1948, p. 18.
2. "UNESCO—What It Is, What It Does, How It Works," Unesco House, Paris 16e, France (a typed memorandum), and other sources.
3. A memorandum from Charles E. Rogers, Information Division, Food and Agricultural Organization, Washington, D.C., February 15, 1948.
4. Memorandum from the Preparatory Commission for IRO, Washington, D.C., November 21, 1947; and a telephone conversation with IRO, Lake Success, N. Y.
5. Letter of February 27, 1948 from Leslie Atkins, U.N. Relief and Rehabilitation Administration, Washington, D.C.
6. New York *Times*, February 8, 1948; *Senior Scholastic*, January 19, 1948, p. 12; Brock Chisholm: "WHO: A Progress Report," *United Nations Weekly Bulletin*, November 18, 1947.

Chapter 14

1. Albert Einstein: "Atomic War or Peace," *Atlantic Monthly*, November, 1947, pp. 31, 32.

References

2. World Government News, December, 1947, pp. 1, 2; letter from United World Federalists, Inc., August 15, 1947.
3. Letter from the Citizens Committee for United Nations Reform, Inc., November 14, 1947.
4. New York Times, July 24, 1947, p. 6.
5. Philip Morrison and Robert R. Wilson: "Half a World . . . and None: Partial World Government Criticized," Bulletin of the Atomic Scientists, July, 1947, p. 181.
6. Letter from Freedom and Union, November 14, 1947, 700 Ninth Street, Washington 1, D.C.
7. Common Cause (975 E. 60th Street, Chicago) July, 1947, pp. 1, 2; World Government News, March 1948, p. 11.
8. Letters from Mitchell Lewin, World Republic, Inc., June 23 and July 23, 1947.
9. Student Federalist, March, 1948, pp. 2, 4.
10. Common Cause, op. cit., p. 2.

Chapter 15

1. "A Library Presents the Atomic Age," 8-page circular issued by the National Committee on Atomic Information.
2. Sally Cartwright: "Where the Atomic Bomb was Born," Progressive Education, October, 1946, pp. 4-6, 43, 44; and letter of August 19, 1947, and memorandum from Philip E. Kennedy, High School, Oak Ridge, Tennessee.
3. A letter of April 21, 1947, and program from Mary Stamps White, Vice Chairman, Atomic Energy Week, Charlottesville, Virginia.
4. Information provided by the Emergency Committee of Atomic Scientists, December, 1947.

Chapter 16

1. Christian Century, January 8, 1947, p. 61.
2. Time, August 18, 1947, p. 64.
3. Otto Zoff: "Young Hands Across the Sea," Coronet, March, 1947, pp. 9, 10.
4. Scholastic, March 1, 1948, p. 12T.
5. Time, August 18, 1947, pp. 77.
6. Memorandum from Susan Fleisher, 224 E. Church Road, Elkins Park, Philadelphia, Pennsylvania, Summer, 1947.
7. New York Times, November 2, 1947, editorial page.
8. Printed Matter from the International Committee of the Y.M.C.A. received in November, 1947; Frank C. Laubach: Prayer, Fleming H. Revell Company, p. 27.
9. Letter from Suzanne Macpherson, Experiment in International Living, Inc., August 27, 1947.
10. New York Times, March 18, 1947, p. 12.
11. New York Times, October 25, 1947, p. 3.
12. Letter from Drew Pearson's office; Christian Century, March 17, 1948, p. 296.

13. George Révay: "Swiss Aid for Europe's Starving Children," *Reader's Digest*, February, 1947, condensed from the *Christian Century*, January 8, 1947.
14. Ruth Leonard: "Village of Children," *Freedom and Union*, July, 1947, p. 10.
15. *New York Times*, August 31, 1947, editorial page.

CHAPTER 17

1. *Collier's*, January 10, 1948, p. 66.
2. Ibid.
3. *New York Times*, January 8, 1948, p. 1.
4. *New York Times*, January 6, 1948, p. 22.
5. *Collier's*, op. cit.
6. *Freedom & Union*, January, 1948, p. 20.
7. *New York Times*, November 11, 1947, p. 14.
8. *New York Times*, March 30, 1948, p. 1.
9. *New York Herald Tribune*, April 3, 1948, pp. 1, 24; *New York Times*, April 4, 1948, Editorial page.
10. *Christian Century*, January, 28, 1948, p. 99.
11. *Changing World*, January, 1948, p. 3.
12. *New York Herald Tribune*, February 9, 1948.
13. $5,300,000,000 was appropriated, not including supplementary amounts for military and economic aid, and 3 per cent of $197 billions (*New York Times Magazine*, January 25, 1948, p. 13) is $5,910,000,000.

CHAPTER 18

Force and Friendliness

1. *Christian Century*, January 14, 1948, p. 38.
2. *Newsweek*, February 23, 1948, p. 25.
3. *Christian Century*, January 28, 1948, p. 99.
4. *U. S. News and World Report*, March 26, 1948, pp. 42, 43.
5. Albert Einstein: "Only Then Shall We Find Courage," Reprint from the *New York Times Magazine* by the Emergency Committee of Atomic Scientists.
6. W. T. Holliday: "Our Final Choice," *Readers Digest*, January, 1948, p. 2.
7. Norman Cousins: "Modern Man Is Obsolete," The Viking Press, New York, 1945.
8. Elton Trueblood: "Alternative to Futility," Harper and Brothers, New York, 1948, pp. 36, 49.
9. *New York Times*, January 27, 1948, p. 6.
10. Norman Cousins: Op. cit., pp. 44, 45.
11. John Foster Dulles: address before the Foreign Policy Association, January 17, 1948, p. 10.
12. *Time*, March 8, 1948, p. 19.
13. *New York Herald Tribune*, February 25, 1948, p. 25.
14. *New York Times*, December 18, 1947, Article by A. M. Rosenthal.

References

Chapter 19

1. "While the Patient Sinks," *Christian Century*, May 14, 1947, p. 615.
2. See Chap. 4.
3. Cord Meyer, Jr.: "What Price Preparedness?" *Atlantic Monthly*, June, 1947, pp. 27-33.
4. *New York Times Magazine*, January 25, 1948, p. 13.
5. Letters from Donald B. McPhail, Bureau of the Budget, Washington, D.C., August 27, September 10, October 17, 1947, and accompanying charts.
6. *New York Times*, September 2, 1947, p. 11.
7. *New York Times*, June 12, 1947, editorial page, quotation from Chancellor Harry Woodburn Chase.
8. Albert Einstein: "Atomic War or Peace," *Atlantic Monthly*, 1947, p. 32.
9. *New York Times*, January 18, 1948, p. 5.

Quotations Preceding Parts II and III

1. Address by General Omar N. Bradley, Drake Hotel, Chicago, May 15, 1947. Released by the Veterans Administration.
2. *New York Times Magazine*, September 9, 1945, p. 13.

Figures 1 to 4

1. Chart "Revenues and Expenditures, Federal Government, Fiscal year ending June 30, 1948 (Revised August, 1947)" shows the following costs: National Defense $10.4 billions, Veterans Programs $7.5 billions, Interest on Debt $5.1 billions, "International" $4.3 billions. Letter from Donald B. MacPhail, Bureau of the Budget, October 17, 1947, indicates that of the $4.3 billions, only $3.9 should be charged to war. This gives a total of $26.9 billions per year, $73.7 millions per day and $3 millions per hour for a 24-hour day.
2. Letter from R. W. Berry, Assistant to the Secretary of Defense, Washington, March 24, 1948.
3. Income: *New York Times Magazine*, January 25, 1948, p. 13. Cost of defense: (see reference 1). Marshall Plan average: $17 billions requested by the administration divided by 51 months (the period indicated) gives $333 millions per month or $3,996 millions ($4 billions) per year.
4. Cost of World War I, lasting 584 days or 1.6 years, was $22,625,000,000—*Encyclopedia Americana*, Vol. 28, 1946, p. 650; average cost per year was therefore $14,140,000,000 ($14.1 billions). Cost of defense: Chart "Revenues and Expenditures" (see above) $10.4 billions plus $320 millions for Greek-Turkish aid. Letter from Donald B. MacPhail (see above), gives a total of $10.7 billions.

Index

Index*

Aircraft, 26f
Airplanes, 37f
Air Policy Commission, 26
Alaska, 49
Alsop, Joseph and Stewart, 49, 156
American Association of Scientific Workers, 41
American Friends Service Committee, 99, 139, 153
American Legion, 120
American Library Association, 129
American Red Cross, 139
American Youth for World Youth, 139
Amidon, Beulah, 9
Anti-Semitism, 12
Appropriations, see Budgets
Arctic Circle, 49, 70
Argentina, 12
Argonne National Laboratory, 83
Army and Navy Munitions Board, 47
Arnall, Ellis, 12f
Arthur, William B., 48f
Articles of Confederation, 115
Asia, 11
Atomic age, 4
Atomic bomb, 5, 24, 31f, 58
Atomic energy, 7, 82f
Atomic radiation, 38

Atomic research, 84f, see also Research, military
Atomic Scientists of Chicago, 34
Austria, 144, 146
Ayer, Coburn, 131

Baba Yar Ravine, 19
Baldwin, Hanson W., 25
Balkans, 149
Bataan, 93
Bevin, Ernest, 143
Bidault, Georges, 143
Biological warfare, 40
Bonnet, Henri, 135, 136
Botulism, 40
Bowles, Chester, 9
Bradley, Omar N., 54
Bubonic plague, 40
Budgets, military, 8, 24f, 162
Bund, German, 12
Burnham, James, 27f

Campbell, John W., 87
Cannon, Walter B., 69
CARE, 167
Carnegie Corporation, 98
Charlottesville, Va., 132
Children, 9
China, 8, 11, 24, 68, 89, 146, 149, 153

* Names of various persons, organizations and publications not appearing in the index may be found in the preface and in the appendices.

Index

Cholera, 40
Christian Rural Overseas Program, 141
Churchill, Winston, 20, 95
Civilian Conservation Corps, 70
Clark, Grenville, 126
Clemenceau, Georges, 95
Coale, Ansley C., 39
Columbia University, 42
Committee to Frame a World Constitution, 125
Communism, 13f, 17f, 74f, 94, see also Russia
Compton, Arthur H., 3
Congress, United States, 95
Conscription of labor, 163
Continental Congress, 115
Corregidor, 92
Cousins, Norman, 128, 151, 152

Dardanelles, 126
Dartmouth College, 97
Dean, Vera M., 128
Debt, federal, 41
Decentralization of industry, 27, 163
Democracy, 14f, 28, 169
Diphtheria toxin, 40
Disease, prevention and cure, 82f; war against, 70
Displaced persons, 9
Draft, 25
Dulles, John Foster, 150, 154, 168
Dunkerque, France, 135
Dunkirk, New York, 135, 167
Duvall, Sylvanus, 72
Dyer, R. E., 85
Dysentery, 40

Education, American system, 57, 58; for public service, 96f
Einstein, Albert, 33, 72, 118, 151, 166
Eisenhower, Dwight, 44
Emergency Committee of Atomic Scientists, 33
Empire, American, 28
Employment, 15
Enoch Pratt Free Library, 127
Essex (aircraft carrier), 42

Europe, 13f
European Recovey Program, 25, 144
Experiment in International Living, 140

Fascism, 12, 13
Federal Council of Churches, 147
Federal income, personal annual, 163
Federal Union of the Free, 123
Federal World Government, 118f
Federation of American Scientists, 164
Fermi, Enrico, 3, 86
Finland, 139
Finletter, Thomas K., 26, 34f
Fischer, John, 17f, 74f, 123, 153, 168
Fosdick, Raymond B., 60f
France, 10, 40, 136, 144, 146, 149
Frank, Laurence K., 69
Fulbright Act, 100, 101

Geneva, 124
Germany, 11, 12, 149, 150
Great Britain, 10, 40
Greece, 8, 11, 24, 147, 149
Greenaway, Emerson, 127
Guerrillas, 11

Hahn, Paul F., 85
Hailey, Foster, 11
Hancock, John M., 50
Harvard University, 138, 143
Hauser, Ernest O., 13
Henri Dunant Center of "Relief for Children," 141
Higinbotham, William, 128
Hiroshima, 31f, 38
Hitler, Adolf, 51, 72
Hogness, Thorfin, 86
Holland, Kenneth, 99
Holliday, W. T., 151
Hoover, Herbert, 76, 153
Hughes, Thomas, 131
Hungary, 10
Hunger, 9f
Hutchins, Robert M., 125

Index

Incendiary bombs, 38
India, 11, 68, 89, 153
Indo-China, 149
Industrial mobilization, 47
Intelligence Service, 25
International Children's Emergency Fund, 109
International Ladies Garment Workers Union, 140
Isotopes, 84f
Italy, 10, 12, 144, 146, 149

Japan, 11
Jefferson, Thomas, 96
Johnson, Senator, 59
Jordan River Authority, 76, 153

Kilgore, Senator, 62
Kingdon, Frank, 130
Kistiakowsky, George B., 6
Korea, 149
Kremenchug, 19
Kremlin, 17, 75, 76, 150
Ku Klux Klan, 12f

LaFollette, Robert M. Jr., 96
Langmuir, Irving, 37, 59, 60
Lawrence, David, 150
Laurence, William L., 3f
Lenin, Nikolai, 74
Lilienthal, David E., 87f
Lindley, Ernest K., 150
Lippmann, Walter, 10
London, 12
Los Alamos, 4
Luce, Clare Boothe, 130

Madison, James, 96
Marshall, George, 150, 154, 156, 162
Marshall Plan, 143f
Marxism, 19, see also Communism
McCauliff, John, 136
McCutcheon, Lieut. Colonel, 38
McNaughton, A. G. L., 156
Medical and Surgical Relief Committee, 140
Middle East Oil, 25, 126
Middleton, Drew, 12

Miller, Jerry, 131
Minsk, 18
Missiles, guided, 38
Molotov, V. M., 144
Morrison, Philip, 32f, 42, 122
Muller, Edwin, 115f
Murphy, Gardner, 68f

Nagasaki, 32f
Nash, Walter, 69
National Advisory Committee for Aeronautics, 38
National Association for the Advancement of Colored People, 15
National Inventors Council, 46
Nationalism, 58
Natural sciences, 60f
Naval bases, 25
Nazism, 12
Negroes, 13, 15
Neutrons, 4f
Newsweek, 85
New York Times, 25
NKVD, 17
Norway, 99, 141

Oak Ridge, Tenn., 82f, 85, 132, 168
Osborn, Frederick H., 146
Osusky, Stefan, 75

Pakistan, 11
Palestine, 8, 149
Paris, 9
Parker, David B., 39
Peace organizations, 66
Pearl Harbor, 48, 68
Pearson, Drew, 140
Peterson, Houston, 70
Plutonium, 83f
Poison gas, 39
Poland, 10, 136
Police, secret or political, 8
Politburo, 18
Poltava, 19
Potatoes, dumped, 14
Power plants, 82f
Prague, 13
Prisoners of war, 8
Propaganda, Soviet, 14

Index

Psychology of war, 66f
Public Affairs Committee, 72
Public Opinion Research, 56

Quota Force Plan, 120

Refugees, 9
Rentschler, Harvey C., 3
Research, military, 26f, 47
Reynolds, Quentin, 135f
Robinson, Donald B., 47f
Roosevelt, Franklin D., 70, 95
Roosevelt, Theodore, 96
Royall, Kenneth C., 48
Royce, Josiah, 2
Russia, 13f, 17f, 28, 40, 46f, 62, 74f, 89, 92, 118, 121, 150, 151, 154, 155, 156, 159, 162, 165

Save the Children Federation, 138
Sciences, natural and social, 60f
Segregation, of negroes, 15
Sermon on the Mount, 157
Shils, Edward A., 34f
Siberia, 49
Silverman, Milton, 38
Silver Shirts, 12
Smith-Mundt Act, 100, 101
Social democracy, 16
Social Science Research Council, 39
Social sciences, 60f
South America, 12, 26, 89, 100
Soviet Republic, see also Russia, Communism and Kremlin
Stalin, Josef, 20, 74
Starvation, 9f
State Department, Washington, 95, 99
Stowe, Leland, 12, 55f, 128
Streit, Clarence K., 123, 130
Student Federalists, 98, 129
Survey Graphic, 9
Swiss Red Cross, 141
Szilard, Leo, 3, 86

Tabun, 39
Teller, Edward, 27, 37, 86
Thimann, Kenneth V., 40
Thompson, Dorothy, 130

Tigris and Euphrates River Authority, 76, 153
Time, 46, 85
Tokyo, 38
Truman, Harry S., 26, 86
Turkey, 24, 147

Ukraine, 19
Underground sites (for military industry), 27, 48
Union Now, 130
United Nations, 57, 95, 98, 99, 105f, 153, 154, 169; Charter Revision, 118f, 160; Food and Agriculture Organization, 112; International Children's Fund, 109, 147; International Refugee Organization, 113; U. N. Educational, Scientific and Cultural Organization, 112; World Health Organization, 113
United Nations World, 39
United States, 14f, 17f, 55f
United States Atomic Energy Commission, 83f
United World Federalists, 119f, 129, 152
Universal military training, 25f, 162
University of Chicago, 50, 86
Uranium, 3f, 83f
Urey, Harold C., 34f, 50f, 60, 86, 120f, 152
Usborne, Henry, 124

War, preventive, 150
Warren, Shields, 86
Warsaw, 9, 11
Wendt, Gerald, 41
Whitehead, Alfred North, 151
White Shirts, 12
Willkie, Wendell, 140
Wilson, Robert R., 122
Wilson, Woodrow, 95
Wofford, Harris, 129f
World Constitutional Convention, 124f
World Government, 115f

World Republic, Inc., 124
"World School," Riverdale, 97
World War I and II, 41f, 67, 154
World War III, 28, 50, 160, 165
World Youth Festival, 13, 15
Wright, Quincy, 68

Yale University students, 99
Y. M. C. A. International Committee, 140
Youth Council on the Atomic Crisis, 98, 132
Youth Hostel group, 99